Keys, Keys, Keys

A Visual Celebration of Amateur Radio's Favorite Accessory

By Dave Ingram, K4TWJ

CQ Communications, Inc.

First Edition
Third Printing

Copyright © 1991 by Dave Ingram, Birmingham, Alabama

Reproduction or use, without express permission, of editorial or pictorial content, in any manner, is prohibited. While every precaution has been taken in the preparation of this book, the author and publisher assume no responsibility for errors or omissions. Neither is any liability assumed for damages resulting from the use of the information contained herein.

Library of Congress Card Number 91-071513

ISBN 0-943016-02-9

Cover photo by Joe Veras, N4QB
Illustrations by Hal Keith
Layout and design by K&S Graphics

Published by CQ Communications, Inc.
76 North Broadway
Hicksville, New York 11801 USA

Printed in the United States of America

Table of Contents

Introduction   vii

Chapter 1. Famous Name Bugs   1

The Vibroplex Saga • The Original, Double Lever, and Model X • The Presentation, Original, and Junior • The Upright, or Vertical, Vibroplex • The Midget • The Model 4 and The Blue Racer • The Lightning Bug • Look-alikes and the J-36 • The Champion • The Zephyr • Martin's Rotoplex • Those Famous Mac-Keys • The Standard and Deluxe Mac-Keys • The Speedstream • Late Model Mac-Key • The Ever-Popular Speed-X • Les Logan's Model 501 Speed-X • E. F. Johnson and Other Model Speed-X Bugs • The Delightful Dows • The Rotatable-Yoke Dow • The Tilted-Yoke Dow • Summary

Chapter 2. Limited Production Bugs   25

The Bunnell Gold Bug • The Bunnell Double Action Key • Breedlove's Codetrol • The Eddystone Bug • The Electric Specialty Kit Bug • The Go Devil • The Mecograph • The Swallow Bug • The Magnificent Mellihand Valiant • The 73 Bug • The Rare Rotating Wheel Bug • Homebrew Bugs • Japanese Bugs

Chapter 3. Collecting, Restoring, Adjusting, and Using Keys   39

Cost Factors • Refurbishing • Adjusting and Using a Bug

Chapter 4. Hand Key Haven   45

The Clapp Eastham Spark Key • Signal Electric's Number 2 Key • The Marvelous Marconi 365 • Your First Key (J-38) • The Pole Changer's Key • U.S. Army Key • British Army Key • U.S. Army Key • Japanese Civilian Key • Hi-Mound HK1 Key • Russian Hand Key • German Keys • Swedish Key • Netherlands Key • Clothespin Key • Camelback Key • Summary

Chapter 5. Telegraph and Code Apparatus of Yesteryear   57

Samuel Morse's First Telegraph • Landline Telegraphy and the Western Union • Continental Versus International Morse Code • Telegraph Sounders • A Sounder Adapter for Modern Transceivers

Chapter 6. Specialty Items   63

Miniature Keys • Special Fingerpieces • Homebrew Electronic Paddle • Gold-Plated Key • Crystal Key • Wild Woody WARC Key • Windscreens for Bugs and Paddles • Vibroplex Goodies • Summary

Chapter 7. Modern Keys and Paddles 73
Vibroplex Paddles and Keys • Bencher Paddles • MFJ/Bencher Combination • G4ZPY Keys and Paddles • Schurr Keys and Paddles • Kent Keys and Paddles • The Legendary Swedish Key • Hi-Mound Keys • The Kenpro Key • Summary

Chapter 8. Classic Rigs For Your Classic Keys 85
A 1930s-Style Hartley Transmitter • A Mating 30-30 Receiver • The Famous 6L6 Transmitter

Index 95

Introduction

This book is dedicated to the thousands of radio amateurs worldwide with a special interest in CW communications and telegraph keys. It contains views of and information on semi-automatic keys or "bugs" of all types, hand or "pump" keys, telegraph sounders, miniature "spy keys," custom fingerpieces, new-style electronic paddles, and much more. Some of our featured items are quite rare, some are more familiar, some are readily available on today's market, and all are absolute winners.

The popularity of collecting and using classic keys and bugs is increasing almost daily and with good reason. These delightful gems speak an international language. Every old key is a tangible piece of history you can hold in your hand, and newer keys are just plain fun to use. Those facts, plus numerous hints on collecting, refurbishing, and enjoying keys, are explained in greater detail in the following chapters. As an additional treat, our closing chapter highlights classic rigs you can build at home and use with your classic keys. We aim to please everyone!

Assembling a book of this nature was a monumental endeavor, and it also entailed bringing together photographs and information from around the world. I wish to recognize and thank the following friends and collectors for sharing their photos: Bill Holly, K1BH; Brad Wilson, KA1GDG; Tony Isch, W2GDV; Alan Dorhoffer, K2EEK; Charles Tryor, N4LMY; Neal McEwen, K5RW; Dick Randall, AD9E; Jim Aguirre, WB7DHC; Bill Dunbar, AD9E; Steve Wilson, K0JW; Rick Van Krugel, VE7FOU; Shige Kawasaki, JN1GAD; Gordon Crowhurst, G4ZPY; and Klaus Gramowski, DL7NS. Our special thanks also to professional photographer Joe Veras, N4QB, for assisting with K5RW pictures and this book's cover; Rose Verlander for typing, proofing, and grammatically critiquing the manuscript; Gail Schieber for editing and coordination of the production of the book; plus Dick Ross, K2MGA, and XYL Sandy Ingram, WB4OEE, for their encouragement to write (and complete!) this book. It was a massive project, but I am sure you will agree the results were worth the effort.

Good luck collecting and using keys, and I look forward to contacting you on 30, 20, 17, or 10 meters in the near future. 73, Dave Ingram, K4TWJ

THE EASIEST-TO-USE RADIO KEYS IN THE WORLD!

AMAZING NEW VIBROPLEX
Super DeLuxe Model
"PRESENTATION"
With Super-Speed Control Main Spring and Other Great New Features

VIBROPLEX
Reg. Trade Marks: Vibroplex, Lightning Bug, Bug

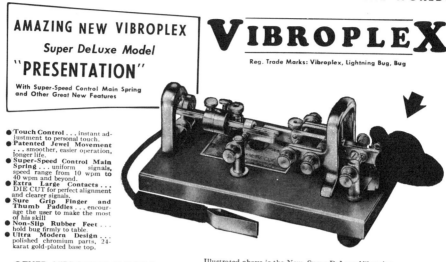

- **Touch Control** . . . instant adjustment to personal touch.
- **Patented Jewel Movement** . . . smoother, easier operation, longer life.
- **Super-Speed Control Main Spring** . . . uniform signals, speed range from 10 wpm to 40 wpm and beyond.
- **Extra Large Contacts** . . . DIE CUT for perfect alignment and clearer signals.
- **Sure Grip Finger and Thumb Paddles** . . . encourage the user to make the most of his skill
- **Non-Slip Rubber Feet** . . . hold bug firmly to table.
- **Ultra Modern Design** . . . polished chromium parts, 24-karat gold-plated base top.

OTHER VIBROPLEX MODELS

Vibroplex Model ORIGINAL, famous the world over for high-class sending performance and ease of operation. Polished chromium parts and base, **$19.50**. Black base, **$15.95**.

Vibroplex Model BLUE RACER, patterned after the Original Model but smaller in size. Capable of the same high-class sending performance for which the original is famous and just as easy to use. Polished chromium parts and base, **$19.50**. Black base, **$15.95**.

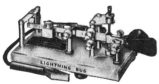

Vibroplex Model LIGHTNING BUG, advanced design, many new features contributing to exceptional sending efficiency and operating ease. Polished chromium parts and base. **$17.50**. Black base, **$13.95**.

Illustrated above is the New, Super DeLuxe Vibroplex Model PRESENTATION, polished chromium parts, 24-K Gold-Plated Base Top.................. **$27.50**

The New, Super DeLuxe Vibroplex Model "PRESENTATION" is the latest word in sending ease and enjoyment. With this amazing New Vibroplex key you'll be able to send better, faster and easier than ever before. So smooth and easy in action . . . strong, firm signals every time . . . suits any hand or any style of sending . . . sets firmly on table . . . built to last a lifetime.

Vibroplex Model CHAMPION, a full size, efficient key designed for radio use only. Modelled after the Lightning Bug and easy to use. An ideal key for the beginner. Chrome finished parts and black base, **$9.95**.

Vibroplex CARRYING CASE, plush-lined, finished in handsome simulated black morocco. Reinforced corners, flexible leather handle. Lock and key. Protects key from dust, dirt and moisture. Insures safe keeping when not in use, **$5.50**.

Write for FREE illustrated catalog and name of nearest dealer.

The "BUG" Trade Mark identifies the Genuine Vibroplex key.

It's your guaranty of complete satisfaction. Accept no substitute.

THE VIBROPLEX CO., INC.
833 Broadway   New York 3, N. Y.
W. W. ALBRIGHT, President

An original-era Vibroplex advertisement.

Chapter 1
Famous Name Bugs

In these opening chapters we spotlight one of the most popular accessories in amateur radio prior to the 1960s—the classic semi-automatic key, or "bug." During its heyday this mechanical marvel was manufactured by a large number of companies, and for several exciting decades it was the true symbol of CW excellence in well-equipped stations everywhere. There were black bugs, chrome bugs, gold bugs, big bugs, and little bugs. There were also bent bugs, tilted bugs, rotary-yoke bugs, and even fully automatic bugs.

This chapter highlights the famous-name bugs most amateurs recognize or remember. Chapter 2 features less well known yet equally attractive bugs. Pull out your pocket magnifier, settle back, and enjoy the tour. The views are great, and the urge to get rolling with your own bug collection will be irresistible!

Every semi-automatic key has its own unique "personality" determined by its vibrating mainspring. This personality is brought to life by its operator. A well-adjusted and smooth-operating bug produces beautiful CW today just as it did years ago, but I must emphasize that practice and technique are most important for effective bug use. Likewise, a properly handled bug is a true complement to its owner. Once you have become accustomed to working with a good bug, you may even slip that old hand key and microphone into a desk drawer. That is when you realize you are really hooked, and it is terrific!

Although today many old-timers still use semi-automatic keys on the air, bug popularity began to decline during the early 1960s. One reason for that decline was the introduction of economical solid-state electronic keyers. Older operators began to lose their fingers' agility, and newcomers never learned how to master a bug. Keyers were easier to use (just bump the appropriate dot/dash levers, and they send perfect Morse), so the evolution was inescapable. We have no criticism in that regard. Most operators sound better on a keyer. Using a bug requires developing a particular skill. If you master it, congratulations! Our hats are off to you!

Before delving into this chapter's views and tales of beautiful bugs, let's point out some of the main components of a bug which we will use for comparisons and discussions. Please refer to Figure 1-1 as we continue.

The main components of a bug are mounted on its base (Figure 1-1[A]). A main yoke (B) with a center pivot rod supports the bug's moving arm (C). Affixed to one end of that arm is the fingerpiece(s) (D), and the other end of that arm is fitted to the bug's mainspring (E). This mainspring is the "heart" of any bug, as it directly determines the semi-automatic key's speed range and "friendliness." A short or stiff spring means fast dots, even

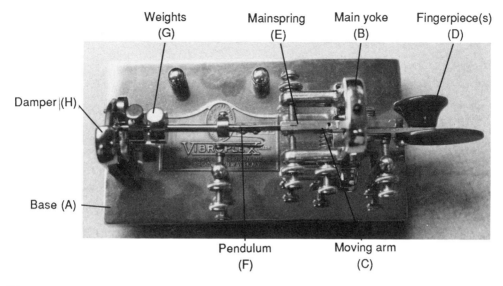

Figure 1-1. The main/working parts of a semi-automatic key, or bug. Components and assemblies shown here will be used for basic descriptions throughout this book.

with extra weights. Some CW operators reportedly filed round grooves in their bugs' mainsprings to achieve slower dots, some used multiple weights, and some ordered two or three replacement arms with affixed mainsprings until they acquired one with the desired action. All too many others simply ignored springs or weights and let the dots fly at 40 wpm while sending dashes at 10 or 12 wpm.

The bug's mainspring (E) is affixed to a pendulum (F) which also moves with dot-making vibrations. One or two weights (G) are set along the pendulum's length for establishing dot speed. Pendulum movement after making a dot or dots is stopped by a rear-mounted damper (H). Other parts include dot and dash contacts, dot/dash travel-stop screws, circuit-closing switch, and so on, and will be explained as we proceed. Meanwhile, let's move on to the stories and illustrations. We begin with a nutshell overview of the all-time biggest name in bugs—Vibroplex.

The Vibroplex Saga

The Vibroplex slogans "The oldest name in amateur radio" and "Old because we are good, not simply good because we are old" are well founded. The company was established in 1890 by J. E. Albright. Bug inventor Horace G. Martin joined the operation in 1904. The original market for Martin's semi-automatic keys was commercial telegraph facilities, but radio amateurs quickly recognized the advantages of bugs and began using them in home setups.

The first Vibroplex keys were made in New York and sported a rectangular nameplate with the words "The Vibroplex by Horace G. Martin." A few years later the nameplate was changed to its presently familiar shape and style, but it did not include the red bug logo. That trademark was added to the nameplate in 1920, and it still appears on Vibroplex items made today.

Bugs carrying the Horace G. Martin nameplate are obviously the pride of collectors, especially if they are in good condition. Some of these classic gems occasionally appear at hamfest fleamarkets or estate sales, but they usually carry a stiff price tag. If you acquire one of Martin's first bugs, try to learn everything possible about its previous owners/history, and document these facts for future reference. Most Martin-nameplate Vibroplexes, incidentally, have survived many years of heavy use and today generally are more suited for display than for actual on-the-air use.

During the years 1904 to approximately 1965 Vibroplex produced over 15 styles, or models, of semi-automatic keys—more than any other manufacturer. Their first model was called the Number One, or Original, Vibroplex, and except for a minor change in the dot contactor, that model continues in production today. It is now made in three styles: the "Standard" wrinkle-painted base version; the "Deluxe" all-chrome version; and the beautiful gold, chrome, and jeweled Presentation version.

In 1911 Martin introduced a second model called the Double Lever. This gem sported dual arms and fingerpieces—one for dots and one for dashes. Before this item was dropped from production (around 1914), a limited number of Double Lever Vibroplexes were also made with a unique cloverleaf design inside their yoke. The Double Lever was neat, but it was a mite difficult to master.

The third Vibroplex was Martin's X model. Introduced in 1912, this bug used a single set of contacts for making both dots and dashes. The pendulum was stabilized by the arm when making dashes, or allowed to move freely for making dots.

Commercial telegraphy was big business during this era, and Martin addressed that market with several interesting bugs. In 1914 he produced the Model 4 Vibroplex, a small semi-automatic key that could easily be carried by telegraphers on special assignments such as boxing matches or ball games. This small bug's name was later changed to the classic Blue Racer, and it was manufactured in standard black base and deluxe chrome with jewel movement versions.

Martin then produced two very unique keys that are top collector's items today: the Martin Upright, or vertical bug (1917), and the extremely rare Midget (1917). Both of these delights used a single pivot post rather than the regular yoke. The Upright was dubbed the "Wirechief's Key" because it occupied minimum space on a busy operator's desk. The Midget was a vest-pocket bug especially designed for traveling telegraphers.

Next came the ever-popular Vibroplex Lightning Bug which sold like crazy and also acquired the military designation "J-36" for use in World War II. This bug was available in standard black, red, green, blue, and chrome-base versions. Thousands remain in shacks today. Martin then produced the Junior Vibroplex (1930), an inexpensive and slightly smaller version of his model Original.

Martin left Vibroplex in 1930, but the saga continued with other models: the Champion; its smaller-base equivalent, the Zephyr; and today's masterpiece of manufacturing, the Presentation. Vibroplex also now makes Vibro Paddles (single and iambic/dual levers) and Brass Racers (paddle or full electronic keyer).

One final and very important note: *Various Vibroplex models are identified by their design, such as yoke, damper, and base differences, not by their serial number.* The key to identifying the model of Vibroplex is visual inspection and photo comparison. Realizing that fact, let's now take a visual tour down memory lane and look at some previously discussed Vibroplex delights.

The Original, Double Lever, and Model X

Our views begin with the terrific trio of Martin's first semi-automatic keys shown in Figure 1-2. They are the Original (upper right), the Double Lever (lower left), and the Model X (middle). These beautiful classics are mounted on Vibroplex's standard $3\frac{1}{2}$ inch wide black base and sport an early "Vibroplex by Horace G. Martin" nameplate discussed previously in this chapter. Notice their "squared off" type yoke and 1910-style fingerpieces. Beautiful indeed! Except for

Figure 1-2. The keys that started the craze! The Vibroplex bugs shown are an Original of 1910-vintage (upper right), a 1912 Double Lever (lower left), and a 1914 Model X (middle). These beautiful classics are in the world-class collection of K5RW.

slight modernization with a more rounded or streamlined yoke, damper arm, and fingerpieces, Vibroplex's Original model is still in production today.

Notice the Double Lever bug has two arms—one for making dashes and another for making dots. A bug with semi-iambic action! Visualize using one of these delights on the air today. Ham heaven for sure, especially if you build a little one-tube Hartley transmitter and regenerative receiver to go with it and add a Tiffany lamp for late-night DXing. I'm serious! I did exactly that, and then I put the rig on 30 meters and had a ball! Best DX? Australia (from Alabama) and with only 5 watts! Vibroplex manufactured the Double Lever from 1911 to 1914 with a regular square yoke, and in limited quantities from 1914 to 1917 with a unique cloverleaf pattern inside its more rounded yoke. The latter is obviously a true collector's item.

The Model X is unique because it uses a single set of contacts to make both dots and dashes. Moving its arm one way closes the contacts while maintaining pressure on the pendulum to inhibit vibration; moving its arm the other way releases the pendulum to make dots. This bug was manufactured from 1912 until 1918 with a flat arm, and from 1918 until 1925 with a round arm. Model X Vibroplex bugs are also a collector's pride.

The Presentation, Original, and Junior

A truly glamorous combination of the old and new is reflected in Vibroplex's presently available Presentation model bug shown in Figure 1-3. This top-of-the-line semi-automatic key features high-luster chrome-plated parts with jewel movement mounted on a gold-plate base. If desired, your name and call letters can also be engraved on this base at the time of purchase.

Presently available also is the Original

Figure 1-3. Vibroplex's gold-base Presentation model semi-automatic key is a top-of-the-line item readily available today. Except for minor changes, it is identical to the Original/Model 1 that Vibroplex began manufacturing in 1904. (Photo via K4TWJ.)

model Vibroplex which looks identical to the Presentation, but lacks the gold-plate base. The Original is also available in a standard/painted base and non-jewel version at an economical price.

The Presentation and Original feature the same basic design as the 1904 Original/Number 1 model bug designed by Martin, and truly reflect the philosophy of difficult to improve upon perfection. The Original (and Presentation) model Vibroplex can be identified by its rounded yoke, round pendulum, and swinging "over the top" damper. The 1948-introduced Presentation is the only bug with an adjustable mainspring (it is secured with screws rather than rivets). Extending the mainspring reduces dot speed range, which is also controlled within that range by weight positioning. Both the Presentation and Original are mounted on Vibroplex's standard 3½ inch wide base.

Another bug introduced in 1934, the Vibroplex Junior, is identical to the Original except for its base. The low-cost Junior is mounted on a trim 3 inch wide base made only in black. It too is a smooth-handling little gem. Only one other Vibroplex, the Zephyr, a small version of the Champion, was made on a 3 inch wide base.

Figure 1-4. The classic vertical Vibroplex, or "Upright" bug. This rare gem was also nicknamed the "Wirechief's Key" because it occupied minimum space on a busy operator's desk. Notice that a single pivot post is used in lieu of a conventional yoke. (Photo courtesy WB7DHC and W7GAQ.)

The Upright, or Vertical, Vibroplex

Your attention (and magnifying glass!) is now directed to the most famous and sought-after Vibroplex of all—the almost priceless Upright, or Vertical, Vibroplex shown in Figures 1-4 and 1-5.

As previously mentioned, this classic bug was also dubbed the "Wirechief's Key" because it occupied a miniscule amount of room on a busy telegrapher's desk. The Upright's design is unusual in several ways. First, a single ball-bearing-equipped, arm-supporting post is employed instead of Vibroplex's traditional

Figure 1-5. Rear view of the vertical Vibroplex showing nametag mounted vertically and circuit-closing lever. Also visible are the right-angle-mounted fingerpieces. This key was made between 1918 and 1925. (Photo courtesy WB7DHC and W7GAQ.)

yoke. Also, a single set of contacts such as in the Model X are used to make both dots and dashes. The nameplate is mounted on the rear. Notice the heavy horseshoe-shaped base with mounting hole.

Several individuals have expressed interest in producing modern replicas of this masterpiece, incidentally, but no finished products have emerged to date. Pity. They would be a real treat to enjoy with a modern transceiver. Maybe someone will at least adapt the vertical design concept to their present-day keyer paddle.

Close investigation of the 1918 to 1925 produced Upright reveals it is actually a miniature bug with angled fingerpieces. As Martin pointed out in his initial description, it could also be placed horizontally and operated like a regular bug. Expanding on that concept, Martin utilized several pieces of the Upright to develop and produce another very special item—the Midget bug.

One other point warrants mention before leaving this story. Another vertical bug was invented by William Coffe and manufactured by the Mecograph Company between 1907 and 1914. Vibroplex acquired Mecograph's patents in 1914, but Martin designed his own vertical bug rather than use the cumbersome Coffe design. It's no wonder, as the Coffe bug looked like a combination of pendulum clock movement and horizontal paddle keyer!

The Midget

Undeniably the rarest semi-automatic key Martin ever made, the Midget, or "pocket bug," is highlighted in Figure 1-6. A very limited number of these little gems were manufactured between 1917 and 1922, and they are so scarce that price-valuing them is impossible. Yes, a genuine Mona Lisa of bugs! Our photograph is compliments of Vibroplex collector Jim Aguirre, WB7DHC, who does not

Figure 1-6. Only two of these very rare and highly prized Vibroplex Midgets are known left in existence today. Notice the "yokeless" construction of this pocket bug and its very narrow base. (Photo courtesy WB7DHC and W7GAQ.)

have a Midget, but shot this picture from friend W7GAQ's collection.

Advertisements for the Midget were run in trade magazines such as *The Railroad Telegrapher* (Figure 1-7). Thus, many amateurs and collectors were not aware of their existence.

The Midget features a single-post arm support like the Upright, but it has a regular vibrating pendulum for making dots, and a miniature "U"-shaped damper similar to the Number 4/Blue Racer. Its base is approximately 2 inches wide by 5½ inches long, and a knurled nut holds its swivel foot. The little tyke weighs only 15 ounces and fits into a vest pocket! I cannot even visualize using one of these gems on the air today, but that is no problem, as no owner would remove one from under lock and key!

The Model 4 and The Blue Racer

We now shift from the rare category to more often found and used Vibroplexes, and begin with the Model 4 and its descendant, the Blue Racer shown in Figure 1-8. Pay close attention to this and following photos and discussions, as they contain a wealth of information on identifying various model Vibroplexes and estimating their time of manufacture.

A variety of changes or evolutions occurred during this little bug's life (1914 to 1966). It came into the world in 1914 with a small yoke, trim "U"-shaped damper assembly, and "squared-off" thumbpiece mounted on a narrow 2½ inch wide base. The standard base was painted Vibroplex's traditional black, and the deluxe version was nickel plated. Upper/working parts of both versions were nickel plated. Around 1922 the base was also made available in blue enamel (obviously a prelude to its name change). Base coloring was again used on the Lightning Bug from 1926 to 1937; red, blue, green, plus standard black and nickel plating were available (more details later). During the 1920s the Model 4 could also be special ordered on a regular size 3½ inch wide base such as that used on the Original, on the later-produced Lightning Bug, and on the Champion (although such a change reminds me of getting a gold medal bronzed!).

All Vibroplex deluxe models, including our sweet little Blue Racer, changed to high-luster chrome bases with chrome upper parts, jewel movement, and red fingerpieces during the 1940s. Standard models also shifted from nickel-plated to chrome upper parts. During World War II all deluxe models changed to painted gray bases. After the war (1945) deluxe bases returned to chrome. Then during the late 1950s all standard-base Vibroplexes changed from black to gray.

The early model Blue Racer is recognized by its narrow (2½ inch wide) base, smaller than Original size yoke, and "U"-shaped damper bar as shown in Figure 1-9. Later model Blue Racers (those made after the mid-1950s) were once again changed to incorporate a small version of the swinging "over the top" damper arm such as that shown in Figure 1-10. Vibroplex discontinued production of this magnificent bug during the 1960s, but it is still a favorite among many amateurs. It is a true beauty, easy to carry, and is fast on dots. Extra weights slow it down to speed, and it is terrific fun to use with modern transceivers. I use a deluxe Blue Racer almost daily and love it. So will you, if you can find one in good condition!

The Lightning Bug

The classic Lightning Bug model Vibroplex, shown in Figure 1-11, requires very little introduction, as it has been used by numerous telegraphers and radio amateurs worldwide. Its unique design and parts combination, however, make it a favored item in any serious key collection. Note the triangular-shaped yoke with round support posts, dual post-supported damper arm, and flat pendulum—a combination exclusive to Vibroplex's Model 6, or Lightning Bug. Two later-produced Vibroplex bugs, the Champion and the

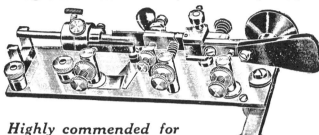

Are You at Your Best?

Martin "MIDGET"
Weight 15 ounces.   Fits vest pocket.   Nickel-plated, $15

To be at his best is the patriotic duty of every Telegraph Operator.

With a Vibroplex, you attain maximum efficiency with a minimum of effort.

The Vibroplex quickly develops ability not possible with the hand, and insures you being at your best.

Highly commended for Radio work

VIBROPLEX No. 4
Blue-Enameled Base - $15
Nickel-plated Base - - $17
(**Mounted on Old Style Base** - - - - $15

Martin Vibroplex

The Vibroplex transmits **STRONG, CLEAR** Morse and carries thru over the longest, heaviest circuits without visible effort on the part of the sender.

A never-ending surprise to Telegraph Operators everywhere, is the EASE with which a Vibroplex is operated.

There's no cramped, tiresome muscular movements in Vibroplex sending. Instead—a smooth, pendulumlike motion of the arm in pressing the lever from side to side—*the machine does the work.*

Why Not Get That Vibroplex To-day?

When you have learned to use a Vibroplex—and it is EASY to master—you have found the QUICKEST and EASIEST way to telegraph.

You do your work in half the time, and with half the labor. Actually pays for itself in labor saved.

Any Telegraph Operator can use the Vibroplex. The old and young use it with equal success. Many have mastered it the first day. What they have done you can do.

You owe it to yourself to take advantage of the EASY ACTING Vibroplex. DO IT RIGHT NOW

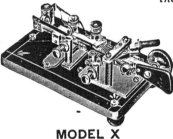

MODEL X
Japanned Base - - - - - $15
Nickel-plated Base   - - - $17

SEND FOR A VIBROPLEX

Immediate shipment—Money order or registered mail. Liberal allowances on old machines.

> To get the best results—Get a Vibroplex for your own personal use. Adjust to suit your hand. Don't tinker. Don't let others use or change the adjustment.

CATALOG ON REQUEST

OLD STYLE SINGLE LEVER
Japanned Base - - - - - $15
Nickel-plated Base   - - - $17

THE VIBROPLEX CO., Inc.
253 Broadway, NEW YORK

J. E. ALBRIGHT, General Mgr., Member N. Y. O. R. T., N. Y. Div. 26. Cert. 68

Figure 1-7. Original-era advertisement featuring the Vibroplex Midget, Model 4/Blue Racer with blue base, plus Model X with pinstriping. (Photo courtesy WB7DHC.)

Figure 1-8. Vibroplex's trim little Model 4 (lower key) and its descendent, the Blue Racer (upper key). See the discussion of this 1914 to 1966 produced classic in the text. (Photo courtesy KA1DGD.)

Figure 1-9. Here is a closer look at the standard-version Vibroplex Blue Racer. Notice its narrow base, round pendulum, and "U"-shaped damper bar. This is a terrific key to use on the air today. (Photo courtesy W2GDV.)

Figure 1-10. Late-model deluxe Blue Racer. This glamorous key was manufactured from the mid 1950s until 1966. Its combination of chrome, red fingerpieces, and jewel movement are breathtaking, and the bug handles great! (K4TWJ collection.)

Figure 1-11. One of the most popular Vibroplex keys ever made—the model 6, or Lightning Bug. Notice the triangular-shaped yoke, flat pendulum, square weight, and flat damper bar. This restored to like new beauty belongs to VE7FOU.

Zephyr, also used a rectangular yoke and flat pendulum, but differed in damper style and base size.

The Lightning Bug was introduced during 1923 with a choice of Vibroplex's standard black base or deluxe nickel-plated base. It became a quite colorful item from 1929 to 1937, when red, green, and blue bases were added to the selection. That is quite interesting, as lightning also exhibits similar colors. Say you've not seen green-sheet lightning? You haven't lived in the deep south; it is synonymous with killer tornadoes.

Like all deluxe-model Vibroplexes, the Lightning Bug changed to a battleship-gray base during World War II, then returned to chrome with jewel movement and red fingerpieces around 1945. A vast number of these famous bugs were manufactured and used by our armed forces during World War II. They acquired the impersonal military designation of J-36. Hundreds became war service souvenirs, and some even appeared for sale in army surplus stores nationwide.

The deluxe Lightning Bug shown in Figure 1-11 was restored to like-new by its proud owner, Rick Van Krugel, VE7FOU. If you follow my *CQ* magazine "World of Ideas" column, which often spotlights classic keys, you know Rick is a master craftsman and machinist. His restoration of keys is equalled only by his handiwork in one-of-a-kind pearl fingerpieces.

Look-Alikes and the J-36

Several independent companies copied the Lightning Bug's design during war years, and look-alikes were also used by the military. One of the more popular copies was Lionel's J-36 shown in Figure 1-12. (Yes, that is Lionel the train company.) Look closely at this key's nameplate, and you will note that it reads "Signal Corps, U.S. Army" (right above "Key Type J-36"). This exceptionally clean bug, incidentally, belongs to fellow collector Steve Wilson, KØJW. Steve has a real knack for uncovering keys and classic rigs neatly packed away for years in original boxes. His secret? Check vehicles as they arrive and unload at hamfests and look under display tables. Be the early bird and offer a fair price.

Figure 1-12. A very popular "look alike" of the Lightning Bug is Lionel's J-36 shown here. This key was manufactured for the armed forces during World War II. (Photo courtesy K0JW.)

The Champion

Vibroplex's low-cost Champion shown in Figure 1-13 is a clean-cut gem that handles as good as it looks. Introduced in 1938, this bug is identical to the standard Lightning Bug, except it uses a single-post damper bar. Both bugs have Vibroplex's standard 3½ inch wide base. In an era when Originals and Lightning Bugs sold for $17 and the fading Junior was $10, the Champion was a hot success at only $9.95! Since cost was paramount, the Champion was only available in a standard/painted-base version. Most of them were gray, but a few were brown.

This delightful bug stayed in the Vibroplex line until the 1980s. In fact, Vibroplex still had a few new Champions for sale as recently as 1988. Maybe there is still one in their stockroom for you. Call them at 1-800-AMATEUR.

Sharp-eyed readers will note KA1GDG's Champion shown in Figure 1-13 is a somewhat scarce left-hand version. This option was also available on other Vibroplexes. Personally, I like using a right-hand bug with my left hand (although I am right handed, I handle keyers or bugs with either hand). I place the key sideways rather than longways in front of me, and

Figure 1-13. The clean-cut and smooth-handling Vibroplex Champion. This bug employs the same triangular yoke as the Lightning Bug, but sports a single damper support post. (Photo courtesy KA1GDG.)

then prop my hand over its top/yoke so my index finger touches the dot knob and my thumb touches the dash oval. I usually send with my right hand, and fill with my left hand when logging, but I've also been known to send simultaneously with a bug in each hand.

The Zephyr

Figure 1-14 shows a neat little gem every CW enthusiast would be proud to own and use: Vibroplex's somewhat rare Zephyr. This bug was manufactured from 1940 until 1948, and it filled the gap left open by the discontinued Junior. Except for its 3 inch wide base, the Zephyr is identical to the Champion. I personally find this size bug perfect. It does not tend to "walk" with heavy use like a little Blue Racer, yet it lacks the massive weight and size of an Original, Lightning Bug, or Champion. Notice the speed-calibrating scale on its pendulum. Views of this delightful bug are compliments of Vibroplex collector and connoisseur Jim Aguirre, WB7DHC.

Martin's Rotoplex

Our visual tour of Vibroplex-related bugs concludes with a key that is not a true Vibroplex, but an item made by H. G. Martin after leaving Vibroplex—the Rotoplex shown in Figures 1-15 and 1-16. Martin produced this item in 1941, and it was used by the Signal Corps during World War II. Study the photos (compliments of collector Tony Isch, W2GDV), and you will see two nameplates. The top one reads "Horace G. Martin, Rotoplex, Jas. Clark Jr. Electric Co., Louisville, Kentucky." The bottom plate reads "Signal Corps—U.S. Army."

The Rotoplex is simple yet effective in design. The arm rides on a large pivoting post similar to a shoulder bolt, which is covered by a metal dome. That unusual yoke is complemented by a dot stop screw also holding a tension-adjustment spring and knurled nut. The damper is also simplified to a basic post holding a thin damper. A creditable number of these bugs were made, but they remain fairly scarce.

Since various models and styles of Vibroplexes may be confusing to new collectors, I have prepared the following simple pocket guide for your assistance. Good hunting and good DXing with your Vibroplex bugs!

Upper parts	3½ inch base	3 inch base	2½ inch base
Standard yoke, swinging top damper	Original	Junior	—
Small yoke, small swinging damper	Special Order Racer	—	late Blue Racer
Small yoke, U-shaped damper	—	—	early Blue Racer
Tripod yoke, flat pendulum	Lightning Bug	—	—
Tripod yoke, flat pendulum, 1 post damper	Champion	Zephyr	—

Those Famous Mac-Keys

Our guided tour of classic bugs continues with another famous name in telegraphy —Theodore Roosevelt McElroy—and his world-renown Mac-Keys. Let's begin with some background information on this world champion of radio telegraphy.

Ted McElroy was a native of Boston, Massachusetts, and he could type an amazing 150 words per minute before graduating from grade school. He became a leading telegrapher for Western Union at the young age of 15, and was well known as the world's champion radio telegrapher during the 1920s and 1930s. McElroy entered his first serious code competition in 1922 and beat all contenders with a copying speed of 56 words per minute—a feat that is still impressive today. McElroy's title was beaten in 1934, but he regained it the following year. Ted then "poured on the steam"

Figure 1-14. The classic Vibroplex Zephyr. This key is very similar to the Champion, except it sports a trim 3 inch wide base. (Photo courtesy WB7DHC.)

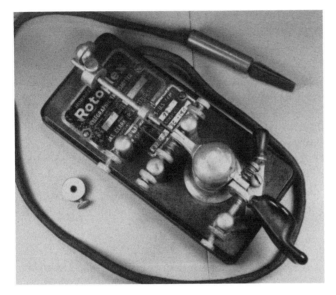

Figure 1-15. Martin's unique Rotoplex. This key has a movable center post with internal bearings rather than a conventional yoke. The design is simple, but the key handles very well. (Photo courtesy W2GDV.)

Figure 1-16. Close-up view of nameplate on the Rotoplex. This shows Martin contracted with Jas. Clark Jr. Electric Co. of Louisville, Kentucky to make the key for the U.S. Signal Corps. (Photo via W2GDV.)

to ensure lifelong possession of the title. He won the last official competition at Asheville, North Carolina with an incredible copy of 77 words per minute. His record still stands unchallenged today.

During the latter 1930s McElroy traveled around the country promoting telegraphic equipment and presenting code-copying demonstrations. He was a true showman and quite a jovial entertainer. One of his favorite stints was to pause in the middle of a high-speed run to drink or smoke, then continue to copy without missing a word. Such ability to "copy in your head" is not unknown today, but remember Ted was also typing from memory while storing additional text in the brain until he caught up!

McElroy began manufacturing his own line of telegraphic equipment in 1934, and it was an immediate success. During World War II his organization produced more telegraphic gear for the armed forces than any other company. They were dedicated and diligent in every effort, finishing contracts ahead of schedule and receiving the Army-Navy "E" award for excellence. McElroy rewarded his employees with blowout parties and jam sessions. Ted sold his company in 1955 and passed on to the great ethereal waves in the sky in 1963. He is survived by a nephew presently living on Long Island, New York.

Ted's famous keys were top-quality instruments several amateurs (including yours truly!) still proudly use today. Their value has increased significantly during recent years, and they now may represent the largest single-item investment in small collections. May Mac's beautiful keys and glamorous tales of origin live forever!

The Standard and Deluxe Mac-Keys

Our visual review of Mac's famous keys begins with Figure 1-17's pair of 1937 delights. The item on the left is a Deluxe model Mac-Key with a simulated marble base. The base was actually heavy metal; special painting created the simulation. The item on the right is a Standard model Mac-Key with a black wrinkle base. Upper assemblies on both models are simi-

Figure 1-17. Golden classics for sure! These deluxe and standard model Mac-Keys handle as well today as they did in 1937, and they are a collector's pride. Both items sport the famous Tee bar of McElroy's early design. (Photo courtesy K5RW and N4QB.)

lar. A Junior model Mac-Key was also available beginning in 1938. It sported the same working/upper parts, but had a stamped steel base. McElroy explained his exclusion of a chrome-base model in a 1938 flyer as follows: "I know as an operator of 25 years experience that chrome creates light reflections that are a severe strain on an operator's eyes." A few years later, however, Mac did produce one model bug with a chrome base—the Speedstream PC-600, or "Teardrop bug."

Notice the use of a unique "Tee bar" rather than a conventional yoke on these early-model Mac-Keys—a feature that immediately distinguishes them from all other bugs. The Tee bar has two significant purposes. First, an owner can easily carry his key to and from assignments simply by curling a couple of fingers around the bar and lifting (see Figure 1-18). Second, the bug can be placed on its side and the pendulum secured with a rubber band for use as a straight/hand key. Notice also the Mac-Key's wide base (4 inches wide, in fact) and thin feet. This combination results in a low center of gravity with plenty of weight and table grip to prevent "walking." Indeed, my Mac-Key holds onto even glass tabletops like a bulldog!

These early Mac-Keys handle beautifully, and they are a sheer joy to use on the air today. They have only one large weight, however, so their minimum dot speed is between 15 and 20 words per minute. An extra weight lowers that speed perfectly, and the keys have a terrific personality.

A close look at the label on the early Mac-Key shown in Figure 1-19 reveals more about its maker. Note "Mfgd. by Theodore R. McElroy World's Champion Radio Telegrapher." Notice also the keys were made in Ted's own hometown of Boston, Massachusetts.

A side view of my Standard model Mac-Key (Figure 1-20) shows its attractive combination of nickel-plated parts and black wrinkle base/Tee bar. This key looks like a true classic and handles like a modern dream! I use it almost daily with old-time rigs and state-of-the-art transceivers alike.

The Speedstream

The later-model Speedstream PC-600 Mac-Key is shown on the left in Figure 1-21 beside a Deluxe Mac-Key. This little critter not only has a rounded yoke and rounded base, but it is all chrome! That was quite a departure from the norm for Ted! The PC-600 in this photo belonged to the famous airline telegrapher Brownie, W3CJI, who used it aboard a DC-3 flying South American routes during the 1940s. Notice the key's feet have been replaced with suction cups for airborne use. This well-known key is now a proud part of one of the largest key collections in existence, that of Neal McEwen, K5RW, near Dallas, Texas. Neal has over 200 keys in several showcases, and all are documented and ready for showing. He requests that visitors call ahead for a tour

Figure 1-18. Author K4TWJ demonstrates how the Mac-Key's Tee bar is used for carrying. (Photo via K4TWJ.)

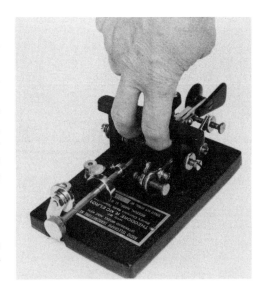

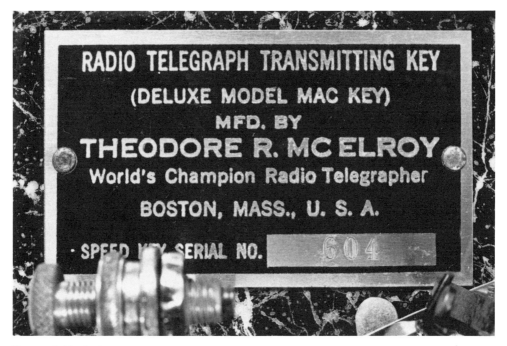

Figure 1-19. Close-up view of nameplate on Mac-Key showing McElroy's title of "World's Champion Radio Telegrapher." McElroy's speed of 77 words-per-minute still stands unequalled today. (Photo via K5RW and N4QB.)

appointment. The collection is in Neal's home. He also works 40-plus hours a week.

Late Model Mac-Key

The post-war model Mac-Key shown in Figure 1-22 reflects changing times and the streamline designs popular during the 1940s. Like curved designs of car bodies, the key's yoke and base are rounded. The famed Tee bar is obviously missing, but "World's Champion Radio Telegrapher" is still on the nameplate. And indeed it should be! Ted's 77 wpm record is unchallenged today, and there is a very good possibility no person will

Figure 1-20. Side view of K4TWJ's Standard model Mac-Key. The bug handles like silk, and it is used daily.

RADIO TELEGRAPH APPARATUS *Manufactured by*
WORLD'S CHAMPION RADIO TELEGRAPHER

NEW SUPER STREAM-SPEED
PLATINUM
Dot Contacts

S-600 PC
$11.85

Into this gorgeous speed key has gone Mac's 30 years operating experience supplemented by the finest engineering ability in the radio-telegraph industry ... with their combined efforts coordinated under the styling genius of one of America's outstanding design artists. See it! Handle it! You'll have to own it! Combining beauty and utility in a most striking fashion, this radically new, semi-automatic key is the last word in operating ease. Fast, rhythmical Morse is a real pleasure with this key.

(A)

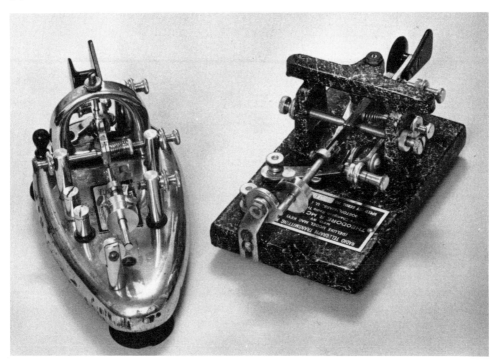

(B)

Figure 1-21. (A) Original-era advertisement for the Speedstream. **(B)** The limited-production Speedstream PC-600 Mac-Key beside a Deluxe model Mac-Key. This particular PC-600 was used by W3CJI aboard a DC-3 flying South American routes during the 1940s. Note the key's feet have been replaced by suction cups for airborne use. (Photo courtesy K5RW and N4QB.)

ever beat it or the famous Mac-Keys. They are great.

The key in this photo belongs to Steve Wilson, K0JW. It is, as all other items in his collection, like new and immaculate.

McElroy also made some straight/hand keys. They will be discussed in another chapter.

17

(A)

(B)

Figure 1-22. (A) Original-era Professional model Mac-Key advertisement. **(B)** Post-war model Mac-Key with rounded yoke rather than famous Tee bar. No radical difference--the key still sings beautiful code! This like-new item belongs to KØJW.

The Ever-Popular Speed-X

If you were a budding young radio amateur with one of those new "KN" callsigns during the 1950s, you surely remember those enticing advertisements of Speed-X bugs that appeared in *CQ* and *QST* magazines. They did not depict great wonders or instant fame, but the keys were quite attractive and had an economical price tag. That did not matter in my case, however, as real money was an unknown luxury. My first amateur radio rigs consisted of junk parts and gratis bell wire wound on a nail form and mounted on a discarded cigar box, but that is a story for another chapter!

Approximately 25 years later I acquired not one, but three (!) Speed-X bugs (good things may come to those who wait, but this was ridiculous!). I now seriously regret never owning a Speed-X during the heyday of bugs. Maybe it is coincidence or maybe it is luck, but all my Speed-X bugs have a very congenial personality, are quite docile, and are easily slowed to straight/hand key speed. In fact, they are my favorite keys for general on-the-air use. If you too have an opportunity to acquire a Speed-X bug, go for it! You will be pleasantly surprised, especially if you enjoy using bugs with modern rigs.

Our story of Speed-X is short and surely incomplete, yet it exemplifies a very noteworthy aspect of key collecting. Every collection is both unique and incomplete, with new information and new items being added on a rather eclectic basis. Collectors learn from each other, trade keys with each other (forget money; what keys do you have to trade?), and expand their horizons on a continuous basis. The only way to write a book on keys is therefore by "taking a picture" at some point in time (the present), relating all available information, and continuing the game from that point. Such is our case with Speed-X. We welcome your input and look forward to including it in a future edition (and giving you well-deserved credit)!

I have uncovered precious little information on the Speed-X operation. It apparently began during the early 1940s under the auspices of Les Logan Co. in San Francisco, California, and shifted to E. F. Johnson Co. in Waseca, Michigan during the late 1940s. The bugs faded from existence during the late 1950s when amateur radio made the transition to electronic keyers and paddles.

Les Logan's Model 501 Speed-X

Surely the most popular version of the Speed-X bug made was this deluxe model 501 produced by Les Logan Co. of San Francisco and shown in Figure 1-23. It is rugged, has a heavy 3½ inch wide base, and handles like silk. Notice the oversized dot/dash contacts; they unscrew from their mount and are replaceable (finding new ones today, however, is nearly impossible!). Two holes in the base are apparently for screwing down the bug on the operating table. I thought they were drilled by a previous owner until I acquired two more Speed-X keys, and they also had mounting holes. As I said earlier, key collecting is a continuous learning process.

Look closely at the dot travel-adjusting screw (on left side of the yoke) and you will see a spring between the yoke's ear and the key's (movable) arm. A knurled nut opposite the one showing (on the travel-adjusting screw) varies tension on this spring, which in turn varies dot arm/

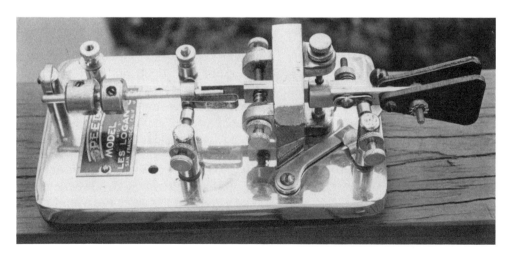

Figure 1-23. Deluxe model Speed-X bug made by Les Logan of San Francisco. Note open-air Tee bar fashioned similar to Mac-Keys. (K4TWJ collection.)

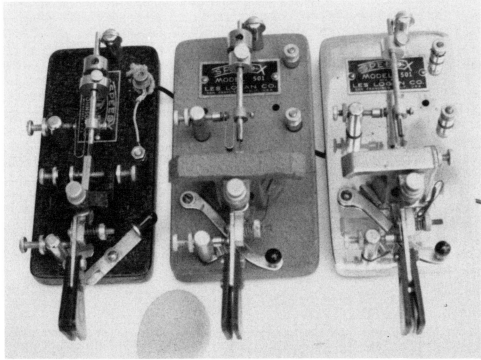

Figure 1-24. (A) Original-era Speed-X advertisement. **(B)** Standard model 501 Speed-X (middle) flanked by the Deluxe 501 (right) and E. F. Johnson's small-based Speed-X (left). The mainsprings in these semi-automatic keys are quite flexible; thus the bugs are very enjoyable to use. (K4TWJ collection.)

fingerpiece tension. Clever! I increased spacing between the triangular fingerpieces on this key with an extra screw and nut (note telltale clue in photo). After becoming accustomed to those flat items rather than a wide Vibroplex-type knob, however, I changed it back to the original. As also mentioned earlier, all my keys see plenty of on-the-air use!

E. F. Johnson and Other Model Speed-X Bugs

In Figure 1-24 the Standard model Speed-X 501 (middle) is flanked by the Deluxe/chrome 501 (right) and E. F. Johnson Speed-X (left). The Standard model sports a clean, battleship-gray base and has the same smooth feel as the Deluxe model. In fact, I find the quality of Speed-X mainsprings surprisingly consistent. If you like one, you like them all! Logan's Speed-X bugs sport a true Tee-bar-molded yoke, and the sides are even formed down for easy carrying. Just hook a finger around each side, lift, and go!

The E. F. Johnson Speed-X is a trim little tyke with a black wrinkle, 3 inch wide base and small yoke (notice size comparison in photo). The famed Tee bar is not included in this model. It also carries Johnson's familiar Viking insignia and model number 114-520 on the label rather than a "500 series" designation. Overall, the bug reminds me of a Speed-X version of Vibroplex's Blue Racer (except its dots can be set much slower). Using this bug on the air is a real thrill comparable only to driving a classic car such as the original Ford "Vicky" or MG-TC. They are terrific!

Although not included in our Speed-X photos, E. F. Johnson Co. manufactured several more Speed-X keys. They are the model 326 hand key, model 515 bug, and model 510 Junior bug. A battery-operated Speed-X buzzer was also available for code practice.

The little Speed-X shown in Figure 1-25 looks identical to the one in Figure 1-24 with one significant difference: The nameplate shows Les Logan rather than E. F. Johnson as the manufacturer! It was apparently made during the sale/transitional period, which, as best we can determine, was between 1945 and 1949. An *ARRL Handbook* for 1949 lists Johnson's ads on Speed-X bugs.

The Delightful Dows

A limited, yet quite fascinating line of bugs was manufactured by the Dow Key Co. in Winnipeg, Canada during the 1950s, and they are special-interest items in many collections today. The unique aspect of Dows is their non-conformity to purely horizontal moving paddles. Their bugs evidently were designed to ease operator strain, but misuse could easily result in a less-than-perfect "fist." One model boasted a rotatable yoke that could be adjusted for right- or left-hand use (or anything in between!). The other model sported a fixed-position yoke set at a 30-degree "uphill" tilt. They were made in black wrinkle, gray, and chrome base versions, plus an all-brass version that polished up to a beautiful luster.

I tried to trace the roots and existence of Dow Key Co. for several years, but uncovered miniscule facts. The key company sold to a specialty company during the 1960s, and a couple of Dows appeared in their store window until the late 1970s. Since then, they have disappeared into the woodwork. Somewhere in the back storeroom of a shop in Winnipeg some beautiful new Dow Keys may still be waiting to be discovered and given a good home. Maybe some of our readers can yet save these classic bugs from extermination—oops, extinction!

The Rotatable-Yoke Dow

Figure 1-26 shows the fabulous rotatable-yoke Dow Key polished up and ready for use. Isn't it wonderful? Would you go crazy over one in your own shack today, and should we plan a scavenger-expedition to Winnipeg to seek and save little lost Dows?

The round yoke rotates inside its fixed outer ring to any desired operating angle, and it is secured with the knurled knob on its left side. Notice everything but the

Figure 1-25. This Speed-X bug looks exactly like the E. F. Johnson item in Figure 1-25, but it sports a Les Logan nameplate. Very interesting indeed!

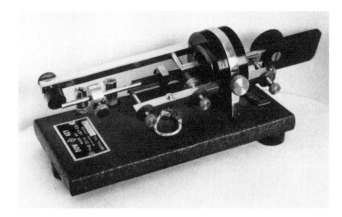

Figure 1-26. The rotatable-yoke Dow. This semi-automatic key can be adjusted for right- or left-hand use, or set horizontally and used as a straight key. (VE7FOU photo.)

base turns with the rotating yoke—dot and dash contacts, travel-adjusting screws, and the long bar holding the pendulum's damper. The key can be set for right-hand operation, swung around 180 degrees for left-hand use, or rotated only 90 degrees and used as a straight/hand key. In the latter case a rubber band or small loop of insulated wire (like that lying on the key's base) holds the pendulum steady to disable dot contacts.

The dot contactor on this Dow Key, incidentally, is a flat strip attached to a square block clamped to the mainspring. This differs from the conventional "U"-shaped contactor typically seen on Vibroplex bugs. Performance, however, is about the same.

The beauty of an adjustable yoke and/or angle of operation has to be experienced first hand to be fully appreciated. You can visualize it by raising one side of your present Vibroplex or Mac-Key. Pure horizontal travel is conventional and nice, but moving the dash arm down at a 30-degree angle (as shown in the photo) really is more comfortable. Surprisingly, the fixed-position/tilted Dow moves in the opposite direction! The rotatable-yoke Dow is truly a remarkable bug and a unique item any amateur would be proud to own!

The Tilted-Yoke Dow

The fixed-position/tilted-yoke Dow Key is shown in Figure 1-27. Place it on a leaning table, and it still stands upright! The tilt was achieved during manufacture by making one side of the yoke shorter than the other. The pendulum is "going uphill"

Figure 1-27. Dow's unique tilted-yoke key. This unusual bug was manufactured with a "lean" to the right for comfortable (?) use. (K4TWJ collection.)

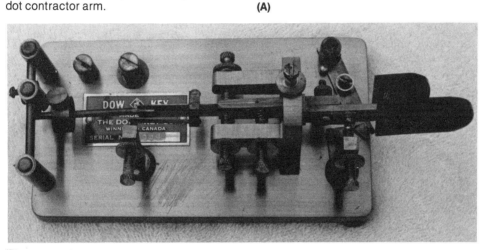

Figure 1-28. (A) Original-era advertisement for a Dow-Key bug. **(B)** Top view of Dow's tilted bug showing nameplate, DK logo, and flat dot contractor arm.

by an approximate 30-degree angle, but it still works fine and makes beautiful dots. The dot and dash contacts also look "bent," but were actually made that way!

Figure 1-28 presents a top view of Dow's tilted bug. Visible in this photo are the "DK" logo, adjustment screws, Lightning-Bug-type damper arm with wheel, and unusual fingerpieces. At first glance, the fingerpieces remind you of rounded end pieces from a pocket ruler separated by a brass spacer. They also come apart in two pieces plus the brass spacer—evidently a Dow Key exclusive! The Dow in this photo is solid brass with such a high luster it looks like gold. I use this key occasionally, but usually on a board tilted 60 degrees in the direction opposite the yoke's tilt.

Summary

The bugs shown and discussed in this chapter were popular items during past eras, and many of them are still used on the air today. Likewise, more information was available on these bugs than on those in the next chapter. It is also interesting to note that bug popularity was greater in the United States than overseas. Most DX operators favored hand or "pump" keys until the advent of electronic keyers. For those who did try to master using bugs, we can only say, "Congratulations and bless you!"

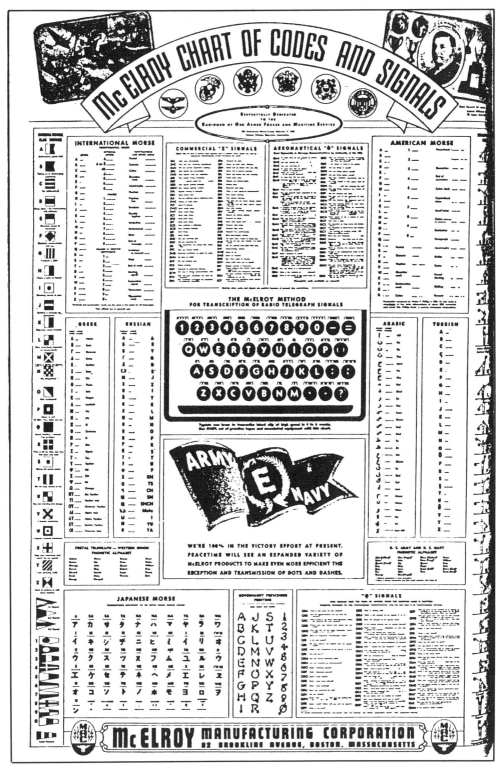

The famous McElroy chart of codes and signals.

Chapter 2
Limited Production Bugs

A significant number of individuals and companies less well known than those discussed in Chapter 1 pursued manufacturing semi-automatic keys during their reign of popularity, and many of the resultant items were quite impressive. That does not mean all their designers reaped a financial fortune in return, however. Some had more mechanical aptitude than marketing saavy, introduced their brainchild when competition was too stiff, or simply joined the game too late for success. Such is the case with many of life's better creations. Backers hesitate to be "the first kid on the block with a new toy" and wait until an existing company does well before following suit.

An interesting number of bugs were also manufactured by foreign companies, and although they were very good keys, those amateurs in countries other than the United States were rather hesitant to accept bugs.

We should also remember that during the early years Horace G. Martin's patents held a tight reign on the semi-automatic key market. Legally-sold bugs had to be different enough in design to be patentable, and that was a formidable challenge at best! Also, several of the successful contenders were quickly bought out by Vibroplex, and their design ideas were utilized to further Vibroplex's prominent place in the market.

Although limited information was available on the keys discussed in this chapter, our hats are off to their designers/ manufacturers for taking the plunge. Many of those hearty innovators would have been absolute winners under the right circumstances. In our own eyes, however, they are all top-of-the-line winners today!

I am sure there are additional items yet unknown to me because the bug market was (still is?) so extensive. If you have information on such keys, please pass it along! We welcome your input, stories, and photos. We will gladly include them in a future edition and give you full credit.

Now let's look at some beautiful special-production bugs that are surely genuine classics!

The Bunnell Gold Bug

Although a well-known and respected name in landline telegraph apparatus and hand keys, Jesse H. Bunnell's involvement in semi-automatic key manufacturing was quite limited. For a short time, however, he did make a beautiful 14-karat-gold bug. We do not have a photo of that rare item, but only an advertisement sketch shown in Figure 2-1. Bunnell's gold 398 featured a very small yoke, flat pendulum, and heavy base. It also sported a single fingerpiece for clean appearance. Although this item was more a trophy or retirement gift than a key for daily use, rumors say it handled exceptionally well. Knowing the quality workmanship of Bunnell, we heartily agree!

No. 398 Gold Bug Automatic Transmitting Key

Figure 2-1. Advertisement sketch of the classic Bunnell gold bug. This item was manufactured for only a short period around the 1930s.

The Bunnell Double Action Key

Although the key shown in Figure 2-2 is not an authentic bug, we are sure you will enjoy viewing and studying its design. This is the famous Double Action Key, or "sideswiper," as it was nicknamed. Many amateurs also dubbed this gem the "cootie key." Notice the main arm is suspended by a short strip of spring metal like a hacksaw blade, and although movement is horizontal, it closes only one set of contacts. In other words, you move the arm right or left of center to make dots or dashes manually—on either side!

Mastering and using this type of key requires skill and practice, but it is indeed possible! I was vividly informed of that fact after writing about the key in my *CQ* magazine "World of Ideas" column. Several amateurs wrote in saying they proudly used sideswipers on the air today, and further investigation revealed they had impressive CW fists! Opinion on use of the Double Action Key varied slightly. Some operators said they mainly made dots with a right/thumb movement and dashes with a left/forefinger movement (such as two left moves, one right move, and one left move for an "F"). Others said they toggled between directions as required (such as one left, one right for two dots, another left for a dash, and another right for the last dot of "F"). Whew!

The Bunnell key shown is solid brass with a black-plastic fingerpiece. Stamped on its right side in small letters is "J. H. Bunnell & Co." This like-new gem belongs to fellow collector and good friend Tony Isch, W2GDV.

Breedlove's Codetrol

One of the most outstanding yet limited-production semi-automatic keys ever made was the unique Codetrol shown in Figure 2-3. This bug was superb in every sense of the word, and it had a great "feel," but something went wrong in marketing. The beautiful Codetrol was made by B. H. Breedlove in Atlanta, Georgia during 1950 and was advertised only a few times in *QST* magazine. The key's right-angle mechanism is fully enclosed in an impressive-looking case, and all adjustments are readily accessible on the front. There is even a calibrated scale (1 to 5) that shows dot speed control.

Only 20 Codetrols were ever made, so they are obviously super-rare. The one in this photo belongs to good friend and master collector Neal McEwen, K5RW. Neal acquired the gem brand-new and in

Figure 2-2. The Bunnell Double Action Key. This item is a "sideswiper" rather than a bug. Its arm moves horizontally, but it closes only one set of contacts. The operator makes dots and dashes manually. (Photo courtesy W2GDV.)

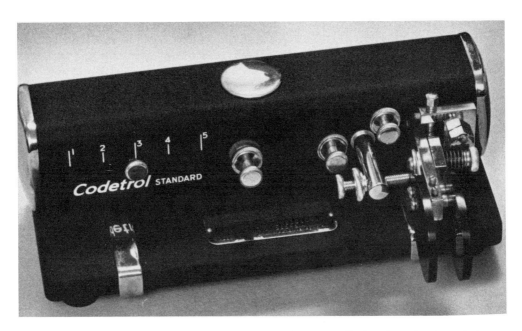

Figure 2-3. B. H. Breedlove made only 20 of these glamorous Codetrol semi-automatic keys during the 1950s. They are very rare and perform great. (Photo via K5RW and N4QB.)

its original box from a chap who overheard him discussing bugs on the air.

The Eddystone Bug

Look . . . on the desk . . . Is it a mouse? Is it a robot vacuum cleaner? Is it an armored beetle? No, it's a classic Eddystone bug! This delightful little critter, shown in formal dress in Figure 2-4, was made in Great Britain from the late 1930s to the early 1940s, and it was truly built to take a whamming and keep on hamming! The key is 8 inches long from tip to fingerpiece, and all working parts are protected by a removable dark-gray cover. The name "Eddystone" is inscribed around this top's hold-down screw, and that shiny dot on the left side is a key-closure switch.

The torpedo-shaped body of this unique bug almost begs to see mobile action. Can you imagine using it with a Gonset Super 6 converter and homebrew CW transmitter in a 1949 chopped-top Merc? Can you hear the whine of that under-the-hood dynamotor now and visualize the bumper-mounted whip knocking hats off gutter-standers while roaring around street corners? Give a cheer if you remember chopped-top Mercs, fuzzy dice on rear-view mirrors, and dynamotors!

Figure 2-4. The unique Eddystone bug. Put three wheels on this little roadster and it will hit 40 words-per-minute. (Photo via VE7FOU.)

The glamorous Eddystone is exposed in Figure 2-5! The working parts of this bug are solid brass and reveal impressive workmanship. First, there is a tripod yoke that looks like a miniature version of the Vibroplex Lightning Bug's yoke. The arm moves on a pivot pin held in position by that yoke's top. Tiny steel balls sandwiched in concave surfaces on the pivot pin give the key a great feel. The damper is a rubber grommet fitted on an "L" bracket that is adjustable within a slotted track at the bug's front. Notice the single and curved fingerpiece—unusual but effective. This key belongs to superb craftsman Rick Van Krugel, VE7FOU, in British Columbia. Rick says it handles beautifully, and he loves it.

The Electric Specialty Kit Bug

Although semi-automatic keys were the rage in telegraphy for approximately 50 years, only one bug was ever available in kit form: the Electric Specialty Company's Speed Bug shown in Figure 2-6. This hearty little key was available for a scant $7.00, and home assembly was a breeze. Mount a few parts on its heavy base, solder wires on under-base terminals, screw on the fingerpieces, and you were ready for big-time CW! These "Cedar Rapids Specials" (so named because Electric Specialty Co. was based in Iowa) were not elite, but they worked surprisingly well for their cost. The key's base and yoke are black wrinkle and its working parts are (lightly) nickel-plated. The rear damper is a large rubber wheel on a slotted support for adjustment. Notice the deluxe model shown in this photo has a circuit closure switch. This like-new key belongs to Steve Wilson, KØJW.

A top view of my own Electric Specialty Kit Bug is shown in Figure 2-7 beside a little Vibroplex Blue Racer for size comparison. Notice this model (standard?) does not have a circuit closure switch. Notice also the nickel plating has worn off several pieces with time. Ignore the missing weights, as I pulled them off for use on another bug (a Speed-X) and forgot to replace them when shooting this photo. I do use the Kit Bug on the air occasionally. It handles fairly well, but dash tension al-

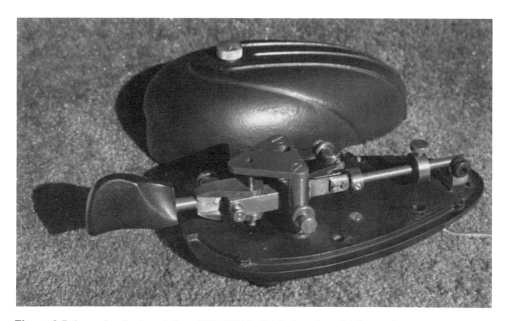

Figure 2-5. An under-the-hood view of VE7FOU's Eddystone bug. Main arm is supported by a tripod yoke with tiny bearings at pivot point. Damper is a rubber grommet on an angle bracket.

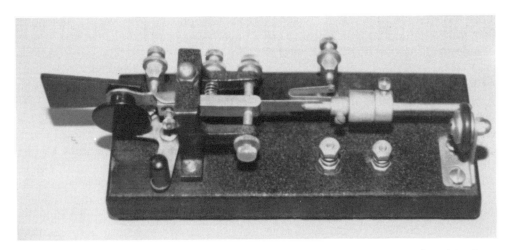

Figure 2-6. The only semi-automatic key ever produced in kit form—Electric Specialty Company's Speed Bug. This key originally sold for a scant $7.00. (KØJW collection.)

ways seems weak. I could restore it to new (or better), but it seems to have more authenticity as is. This key is a perfect mate for those nostalgic basement setups with a Viking II transmitter and BC-348 receiver. Remember lugging home

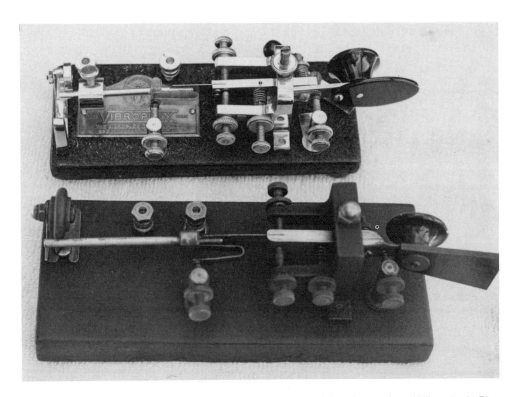

Figure 2-7. Size-comparison of Electric Specialty's Speed Bug (bottom) and Vibroplex's Blue Racer (top). The Speed Bug is approximately 3½ inches wide by 9½ inches long. (K4TWJ collection.)

those monsters in your little red wagon? Remember getting your first real high-voltage shock across the key's contacts?

The Go Devil

A number of creative individuals tried their hands at manufacturing bugs after Horace G. Martin's "exclusive patents" expired, and one such result was the Go Devil shown in Figure 2-8. Approximately 400 of these keys were made by A. H. Emery of Poughkeepsie, New York during the mid 1930s. At least 100 of them are yet unaccounted for and may be waiting to be discovered in out-of-the-way places, probably around New England.

Notice the basic and easy-to-duplicate construction of the Go Devil. The yoke, for example, is made up of a flat horizontal bar screwed to two metal uprights. The thick base appears to be iron or pot metal. The damper is a simple metal rod with a hook. It has a second purpose, however. The rod can be extended so its hook prevents the pendulum from vibrating. Dot contacts still "make," so the Go Devil becomes a sideswiper! Although I have never used a Go Devil (a deprived life, for sure!), I understand its "feel" is basic and plain but effective.

The Mecograph

As we have discussed, the early days of bug manufacturing were almost a private party for their inventor, Horace G. Martin. Competitors had to beat the old boy at his own game, and that was not an easy task!

One of the first semi-automatic keys to defeat Martin's patents was the right-angle bug shown in Figure 2-9. This item was made by the Mecograph Company of Cleveland, Ohio during 1911 or 1912. The key is black with few nickel-plated parts, so details are difficult to see in a photo. The Mecograph's two weights (left side) are fixed; dot speed is adjusted by varying the pendulum's pivot point. Dots are made by releasing tension on the pendulum's spring, and dashes are made manually as usual.

The Mecograph had a short life, since Vibroplex bought out the company in

Figure 2-8. A. H. Emery's famous Go Devil. Approximately 400 of these semi-automatic keys were made during the mid 1930s. (Photo courtesy K5RW and N4QB.)

Figure 2-9. The right-angle Mecograph bug. This item was one of the first semi-automatic keys to defeat H. G. Martin's patents. (Photo via K6ARE.)

1913. It is interesting to note that a few years later Vibroplex introduced a conventional bug that made dots by releasing spring tension.

The Swallow Bug

Names rather than model numbers have always been popular for semi-automatic keys, and every manufacturer strived for originality in that area. One of the resultant unique-named bugs was the Japanese-made Swallow shown in Figure 2-10. This neat key was manufactured by the Dentsuseiki Company in Tokyo during the late 1940s and early 1950s, and it handles beautifully.

The base, yoke, and damper bar of the Swallow are plastic, and all wiring is handled via the chrome-plated strips on each side. Since everything is raised from the base, maintaining the key's like-new condition is a cinch. Notice the nameplate's top edge is calibrated 0 to 9. This serves as a reference for the weight's position in estimating dot speed. The thumb fingerpiece sports a nice curve, and the key is a real pleasure to use on the air. Metal weights are molded into the base, making it heavy like a Vibroplex, and the feet are formed like little stoppers to really hold onto a table or glass surface.

This is the only bug I have successfully used mobile, probably because it is trim (only 3 inches wide), semi-enclosed (a fall

Figure 2-10. The Swallow key was made during the 1950s by Dentsuseiki Co. of Japan, a forerunner of Hi-Mound. Notice all wiring is above the base and via chrome side bars. (K4TWJ collection.)

would not ruin it), and because I can add extra tension to avoid excessive "babble." Mobiling with a bug on bumpy southern roads can be quite challenging. If you try such stints, take my advice and pursue it only on smooth freeways.

We understand that an identical-looking bug was also made in the United States under the Skilman name. I have not seen one of those items yet, but if it performs like the Swallow, it is surely a gem.

As a last note to the Swallow story, after writing this chapter I was graced with a visit by well-known key collector JN1GAD. Shige informed me that Swallow was an early name for Hi-Mound, Japan's largest manufacturer of keys.

The Magnificent Melihan Valiant

Whip out your magnifying glass again because the super-keys shown in Figure 2-11 are really special. These 8 pound beauties are the Deluxe and Standard Melihan Valiants, and they are fully automatic! The keys have two pendulums—one for dots and one for dashes! Look closely at their mainsprings and you will see how their timing is achieved: the dash spring is three times longer than the dot spring. Moving the key's lever to the right causes only the left pendulum to vibrate (fast). Moving the lever to the left causes only the right pendulum to vibrate (slowly). Imagine firing up one of these delights during the CQ World-Wide DX CW Contest. You would definitely have a ball even if you did not win!

These magnificent works of art were introduced by Melvin E. Hansen, W6MFY, in 1939, and they were true classics right from the start. Approximately 500 Valiants were made by Hansen personally or by his assistant, the Schultz Tool and Machine Manufacturing Co. in Anaheim, California, between 1939 and 1950. Most of the glamorous Melihan Valiants were specially made for amateurs desiring a fist and signal like W6MFY's.

Figure 2-11. The magnificent Melihan Valiant. This dual-pendulum marvel is fully automatic. Different length mainsprings are its secret of operation. These beautiful classics were made during the late 1930s by W6MFY. (Photo via K5RW and N4QB.)

Figure 2-12. Proud Valiant owner Neal McEwen, K5RW, demonstrates this magnificent key in operation. There are 17 adjustments on this 8 pound masterpiece. (Photo via N4QB.)

Hansen had a great location for operating, and his antenna was strung high between two oil derricks. His bodacious signal thus drew plenty of attention. Others using his unique key visualized achieving equal results.

Only a small number of Valiants are known to be in collections. Where are the others? Are they packed away in attics or stored with their rigs of yesteryear? We have no magic answers, but if you uncover one (or two), please keep us foremost in mind!

Neal McEwen, K5RW, owner of the beautiful Melihan Valiants shown in Figure 2-12, puts the Deluxe model through its paces. There are 17 adjustments on this mechanical masterpiece, so full "tune ups" are lengthy, but tremendous fun!

The 73 Bug

Everyone loves little bugs, and the gem shown in Figure 2-13 is a real heartthrob! This admirable little bug was designed by a railroad wirechief and manufactured by the Ultimate Transmitter Co. of Los Angeles, California in the mid-1920s. The key was popular among west coast telegraphers because it was small (3½" × 2½"), fully enclosed by a locking cover (removed here for viewing), and easily carried. The 73 Bug was supplied complete with its adjusting tools.

The key's mechanism is quite interesting. Its mainspring and pendulum are mounted right behind the yoke, and its single weight is on the right side. The damper barely clears the cover. This 73 Bug belongs to K5RW. An original-era advertisement for the 73 bug is shown in Figure 2-14.

The Rare Rotating Wheel Bug

We had heard several tales of a very unusual bug that lacked a mainspring or pendulum. Instead, our vague item of unknown origin had a horizontal balanced wheel that turned to make dots. Was it real?

In 1987 Albert Lewis of Nantucket, Massachusetts stepped forth to voice a resounding "yes indeed!" Albert said he actually used one years ago. It did not handle very well, so Albert converted it to a pendulum type which also did not handle well. He threw away the rare gem a few years later. The point, however, is such an item did once exist! Albert did not have any photos of the key, and his hand sketch in Figure 2-15 is all the detail we have on this key—no dates, no names, nothing else. We naturally wanted to share views (notes) of this elusive item with you. Somewhere out there are more details and maybe even a working model.

Homebrew Bugs

While on the subject of sketches and unusual bugs, we must also recognize those dime-store clunkers assembled by amateurs during Depression years. One

Figure 2-13. The ideal bug for travelers—Ultimate Transmitter Company's "73" bug. This item looks like a regular bug with its parts "wrapped around their inside cover." (Photo via K5RW and N4QB.)

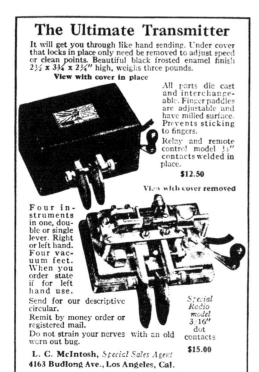

such "Rinko Special" is illustrated in Figure 2-16.

This homebrew bug is assembled on a wood base, and its vibrating spring is a discarded corset spring! The spring is bolted to a metal arm, which in turn moves on a shoulder bolt mounted on a metal strap (shoulder bolts are threaded only half way; the other half is smooth). The metal strip is screwed into the wood block, thus making a yoke. A sewing thimble filled with lead or solder is force-fitted on the spring's "far end" to make a weight. The damper arm, contact supports, and spring supports are small "L" brackets fitted with screws and double nuts. Small springs are squeeze-fitted over travel-limiting screws for dot/dash tensioning. The fingerpiece or "thumbeater" is a piece of wood dowel.

Figure 2-14. Original-era advertisement of the "73" bug.

Figure 2-15. The only known sketch of a rare rotating-wheel bug. Paddle movement causes the wheel to turn and make dots.

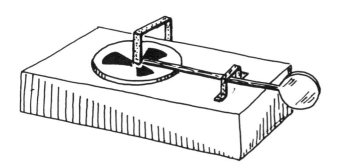
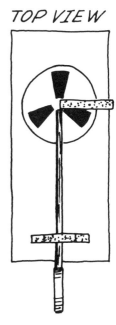

The first impression of this contraption may stop an eight-day clock, but it works! Build one yourself and see!

No corset springs in your junkbox? Don't fret. Figure 2-17 shows another homebrew design using razor blades and rubber bands. Assembly of this big-time "Bandito Delight" begins by sandwiching a thin razor blade between two thin metal strips secured with screws and nuts. The strips are bent or sprung so that a vertically-mounted bolt slips between them to make a pivot point. Rubber bands are then used in lieu of springs for tensioning, and a strip of plastic (like the end of a toothbrush) is used as a fingerpiece. Another strip of metal is bolted to the razor blade's far end for a pendulum, and a large nut is used for a weight. The dot and dash contacts are screws with double nuts mounted on binding posts. File their ends flat and rub a little solder on them for contacts.

If you can set the rubber-band tensions right and hold the key's wood base while

Figure 2-16. Outline of a homebrew bug using a corset spring for its vibrating reed.

sending, this little key will actually work. We do not, however, make any claims of an outstanding "feel"!

Japanese Bugs

We wrap up this chapter with some very impressive semi-automatic keys from the Orient.

First there is the rugged Muse key shown in Figure 2-18. This beautiful little gem sports a round yoke with hefty sidebars and an equally rugged rear damper bar. The white fingerpiece is wide in the

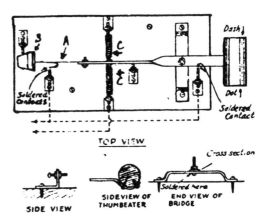

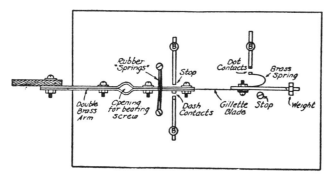

Figure 2-17. Alternate design for homebrew bug. This no-holds-barred item uses a razor blade and rubber bands for gusto CW operation.

Figure 2-18. The Japanese Muse semi-automatic key. Note unique-shaped white fingerpiece and rugged design of yoke. This bug handles very well. (Photo courtesy JN1GAD.)

Figure 2-19. These Japanese "bag" keys bear a close resemblance to United States items and were quite popular in the Orient in the past. Upper key has been modified for left-hand operation. (Photo via JN1GAD.)

middle and tapered on each end. The heavy metal base is 3 inches wide. This key belongs to fellow collector Shige, JN1GAD. Shige did not tell us how it handles, but the key looks like it would be delightful to use on the air today.

The two neat items shown in Figure 2-19 are Japanese "bag keys" model BK-100, also in JN1GAD's collection. The lower bug is a standard right-hand model; the upper one has been modified for left-hand operation.

The design of these bag keys is quite similar to Johnson's clear-top-enclosed bugs of the late 1950s. As such, they should handle very well. Judging by the lip around the BK-100's base, there may even be a clear plastic top for these keys. The plate on each key's left side, incidentally, is marked "speed" and calibrated 0 to 9. The mainspring, pendulum, weight, and damper on these bag keys are quite similar to the Swallow. Although unconfirmed, I hear these attractive bugs are still available in Japan.

Have we inspired or rekindled your interest in semi-automatic keys? Great! They are indeed enjoyable to use, but you must first find one with a good personality and adjust it (and your fist!) to send beautiful code. These subjects are discussed in the next chapter.

Chapter 3
Collecting, Restoring, Adjusting, and Using Keys

This chapter is dedicated to the ever-increasing number of amateurs interested in collecting and using classic keys and bugs. It is short and to the point, but it includes a wealth of information and success-proven notes you will find beneficial. Follow our general guidelines, enjoy restoring and using each newly-acquired item as you go, and key collecting will be a very rewarding experience.

Finding the keys or bugs of your dreams is the challenging part of this game, and personal ingenuity plays a major role in your success. Many newcomers look toward known collectors for "extra keys" to begin their own collection, but such methods are seldom fruitful. Money has miniscule value to collectors. You simply must have attractive keys to trade for other keys. Some golden oldies are occasionally sold just as some classic cars are sold by collectors once in a blue moon, but do not expect bargain prices. Your best bet is combing the countryside and building your own collection of "lovables" and "trade stock."

Another point: Always include a self-addressed, stamped envelope when writing to other collectors and be patient for replies. Most collectors are extremely busy people who are willing to help you, but they are hard pressed for spare time (especially your author)!

There are no cut-and-dried rules on where to look for keys and bugs, but here are some suggestions to guide your search. Every place from attics and basements of old-time amateurs or telegraphers to defunct railroad depots and hamfest fleamarkets is fair game. The latter can be rough picking, however, as more folks are collecting keys daily, and the goodies disappear as fast as they hit a hamfest table. Your best bet is to check with fleamarket traders as they arrive and before they move their items into display areas. Do not get in the way or be overbearing; just ask if they have any keys for sale, offer a fair price without dickering, and keep an open eye for additional key sources. If a trader already has his/her wares out, look closely before moving on. What's in those closed boxes in the background or under the table? Bugs? Wearing a cap or badge stating you buy keys can also prove beneficial, but it can also label you as a sucker for a rip-off. You be the judge.

The want-ad section of amateur radio magazines (United States and overseas), newspapers, and club bulletins are additional good hunting grounds for keys. If there is a telephone number included in the ad, use it! The early bird gets the bug!

Now think. Where else might you find old keys? Visit antique shops and emporiums. Check small towns that may be overlooked by the crowd. Does your local Western Union hold a monthly meeting of retirees? Do you know any railroad telegraphers? Do they know other telegraphers? Opportunities are everywhere. They are both obvious and elusive. Keep an open mind, think positively, and good luck!

One final note: If you can't wait a minute longer to get rolling with your own bug collection, one company can make your dream a reality right now—Vibroplex. You can telephone them toll-free at 1-800-AMATEUR and order an Original Standard, Deluxe, or gold-plated Presentation model bug and get on the air in style.

Cost Factors

I often receive inquiries regarding the value of different keys, and quite honestly, there are no proper answers to such questions. If you are buying, you want to hear a low figure, and if you are selling, you want to hear a high figure. Truthfully, I can only say that any particular key is worth what the buyer will pay and the seller will accept.

A few years ago a fellow collector paid $600 to $700 for a Vertical Vibroplex in good condition. Another friend on the opposite side of the United States purchased a Vertical Vibroplex in similar condition at a public fleamarket for $20. While looking through the Dayton Hamvention fleamarket a few years ago, I found a Mecograph bug in unappealing condition with a price tag of $250. I thought that was steep, until I found another fleamarket dealer with a Standard Mac-Key for sale at $400!

Meanwhile, at another convention another friend bought a beautiful round-yoke Mac-Key for $15, and a fellow collector found a Rotary Dow mixed in with "radio junk" purchased from an old store for $22. As I said, there are no set rules in this game!

If you are fortunate enough to find your dream key for sale, don't procrastinate. Go for it! Let's clarify that point. If you buy a new Original model bug from Vibroplex, it will cost between $70 and $100 (approximately). If the (used) key you are contemplating buying from an individual is in similar (like new) condition, does it not have the same value? Until a few years ago Vibroplex had a few Champion model bugs in stock for sale. The cost was approximately $60. Suppose you postponed that purchase until extra money was in hand. The Champions have now been sold and no more are available. Sixty dollars would be a bargain price today.

Another point: If you have an attractive bug for sale, strive to avoid the "mail-order auction" syndrome of selling to the highest bidder. As an alternative, I suggest originally stating your lowest acceptable bid, and then setting up a conference telephone call so all interested parties can bid on the key directly. Everyone will be more pleased with the results.

Finally, plan your key finances to fit your lifestyle. Most collectors prefer owning many low-cost items rather than few high-cost items, but personal preference is always variable, and no two key collections are alike. Remember those facts.

Refurbishing

If you ask ten different collectors their opinion on key restoration, you will get ten different answers. Some believe in retaining everything in as-found condition, some opt for total rebuilds, and some prefer a happy medium. My only suggestion is this: Do your own thing, but don't destroy the key's original beauty. Avoid painting a worn base in some unusual color. Don't sandpaper chrome parts or contacts, and be careful to avoid breaking fragile old pieces. Take your time on clean up; haste makes waste.

Dirty bugs usually clean up nicely with mild soap and water. Ultrasonic baths are okay if used with care. Remove all parts from the base prior to cleanup. Lay them out according to their sequence of reassembly, and/or draw a sketch for easy reassembly. Screws and washers extending through the base to terminals, contacts, and wiring often short during reassembly. You can sidestep this by slipping short pieces of electrical spaghetti or small cable insulation over the screws before reassembly. If you are refurbishing a Vibroplex, their factory has a fair supply of new replacement parts worth considering. If your bug has a painted base, a light coat of hair spray before remounting upper parts will restore original luster. Chrome pieces can be polished with automobile wax. Excalibur and Mother's Gold wax are superb. Gold can be renewed with a touch of toothpaste hand-rubbed by your finger and rinsed with clear water. Dry all parts thoroughly before reassembly. Beyond that you are on your own!

Adjusting and Using a Bug

Since proper adjustment of a semi-automatic key is very important to smooth operation, let's begin with an overview of that procedure. I will also add some modern-day variations to help both newcomers and old-timers with "stiff fingers" along the way. Please refer to Figure 3-1 as we continue.

First study your bug to ensure its main arm and pendulum look straight in line with the base. Temporarily adjust the left- and right-arm travel screws as necessary. The pendulum's far end should rest lightly against the damper wheel. If your bug's damper wheel support is adjustable, reset it as required. Otherwise, compensate with the right-arm travel screw. After making a string of dots, the pendulum should snap back (lightly) against the damper wheel with just enough tension to stop/damp further pendulum vibration. Check to ensure the weights do not hit or bind on the damper wheel.

Next reset the left-arm travel screw so the dot/thumb fingerpiece moves $\frac{1}{8}$ inch to make dots. Now adjust the dot contact screw (and arm's dot contactor, if it is misaligned) for a solid string of 15 to 20 dots (after which the arm's dot contactor should rest solidly against the dot contact screw to produce a continuous tone in your monitor). Next adjust the left-arm tension spring for your desired tension. Got it? Okay, turn the screw some more and double that tension. Why? You are adjusting a bug, not an electronic keyer paddle. Unless you have a magic telegraphic arm (and trust me, they disap-

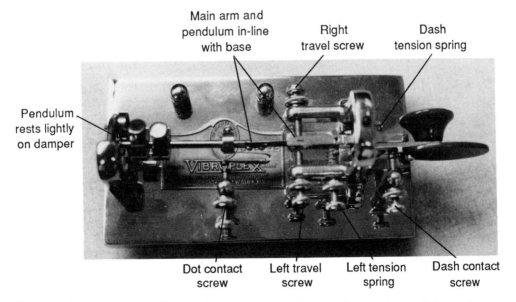

Figure 3-1. Basic point of adjustment on semi-automatic keys. See text for a full discussion.

peared with high-button shoes!), you need to get some real wrist action going to make a bug sound beautiful. The major entanglement today's operators experience with bugs is contacts too closely spaced and not enough tension. They shoot for a feather touch, and then every extra hand or finger movement causes mistakes and sending is choppy due to lack of feel. The additional tension and travel I suggested also brings out the beautiful action of old mainsprings in aged bugs.

Now adjust the dash contact screw so its related fingerpiece also travels $\frac{1}{8}$ inch. Adjust the dash tension spring to equal the dot's tension, and then add a little more dash tension. Do you now hear a pronounced "click-click" on dash make and break? Very good!

Final bug adjustments are best made by connecting the key to a CW monitor or by wiring it in parallel with your electronic paddle, if it can be switched for "tune/continuous keying." Are the bug's dots light or choppy compared to the keyer's dots? Turn the bug's dot contact screw clockwise very slightly. Is the pendulum vibrating after making dots? Reset its right-arm travel screw. Check the arm's pivot screw. Do not overtighten it, however, or the bug's feel will be stiff. Are the dots too fast? Probably! Add a second or third weight on the pendulum to slow the bug to a calm pace (that extra arm travel and tension you added will really be appreciated at this point). Say you cannot find an extra weight? Think back a few pages, and call Vibroplex (1-800-AMA-TEUR). Can't wait? Replace your weight's hold-down screw with a short piece of screw stock and slip a fully assembled PL-259 complete with RG-58 adapter on it. That will cut the wild bug down to size!

The only shortcoming at this point will be achieving smooth dashes and an overall rhythmic sound. These aspects result from practice and use, and that brings us to this chapter's final topic.

Using a bug requires more wrist action than using an electronic paddle, just as using a hand key requires more wrist action than using a bug. The old "straight arm" method (with the key placed beside your rig rather than in front of it) was good for its time, but it does not work for today's amateurs.

Position the bug longways in front of you—between you and your rig (Figure 3-2). Curl your hand around the fingerpieces, and then slide the bug forward so your finger(s) must move back slightly to avoid making a continuous dash. Now try sending perfect code like your keyer sends. Practice continuous dashes—10 to 20 at a time. Remember: The fingers move back rather than pull forward. Now work on commas (– – . . – –). The bug will

Figure 3-2. K4TWJ's easy-to-master technique for enjoyable bug operation. Discussion in text.

Figure 3-3. Operating a right-hand bug with your left hand can be suprisingly easy. Simply curl your left hand over and around the key's top so your index finger rests on the dash knob and your thumb is on the dot fingerpiece.

probably "walk." Hold its base or dash screw with your left hand (no high-voltage shocks with today's solid-state rigs).

Continue practicing, practicing, practicing, and then get on the air and enjoy Morse at its best! Once you get the feel of a semi-automatic key, you will love it. Bugs have a wonderful personality and are capable of beautiful-sounding Morse that cannot be duplicated with a modern electronic keyer.

If you become totally hooked on bugs, work on sending with either hand. Operating a right-hand bug with your left hand is surprisingly easy with a little practice. As demonstrated in Figure 3-3, simply curl your left hand around and over the key's top so your index finger rests on the dash knob and your thumb is on the dot fingerpiece. After perfecting this technique, you can transmit with one hand and log with the other hand during contests. As I mentioned earlier, the enjoyment of collecting keys is equalled only by their actual on-the-air use!

Chapter 4
Hand Key Haven

The items featured in this chapter are synonymous with amateur radio. They are as common to communications systems worldwide as microphones and antennas. They are well known as straight keys, hand keys, and/or pump keys, and they are made in an endless number of shapes and styles. We feel quite confident stating one such manually operated key is an integral part of every enthusiastic amateur's setup.

Hand keys reflect more than amateur radio tradition, however. They are functional items for transcending all language barriers. Hand keys do not require internal or external power sources for operation, they are unaffected by keying voltages or polarities, and they are easy to use. In today's world of beautiful bugs and fancy electronic keyers, hand keys truly receive less than their deserved recognition and respect. Hopefully, the following views will alleviate that dilemma. There are only two kinds of hand keys—good and great!

Before delving into the following views and descriptions, I must point out that no book on keys (including this one) has ever been complete. Please forgive us if we overlooked your favorite key. Send us photos and information about your keys and we will gladly include them in a future edition. Now let's review some very impressive hand keys any amateur would be proud to own and operate!

The Clapp Eastham Spark Key

This chapter begins with reflections of yesteryear and the true Cadillac of spark keys—the famous Clapp Eastham pump key shown in Figure 4-1. This historical showpiece was made during the early 1900s and used by many big-time spark operators. Notice the key's large ½ inch contacts and marble base. These pieces were designed for more than eye appeal; they were necessary and functional. That is because most spark rigs were real fire-breathing dragons, and they were keyed

Figure 4-1. Clapp-Eastham spark key. This item was made during the early 1900s. It sports half-inch contacts for handling high current, and it is the true Cadillac of spark keys. (Photo courtesy K6ARE.)

Figure 4-2. Signal Electric's spark key. The Navy knob on this gem was for more than appearance, as the contacts literally flamed with high current during use. (Photo via W2GDV.)

in their high-voltage transformer's primary line! Current levels were quite high, and small contacts pitted or burned rapidly.

Seeing a spark rig and key in action was a breathtaking experience not easily forgotten. Flames shot from the key, an operator's hair occasionally stood on end, the smell of ozone filled the air, and the antenna glowed purplish-blue against an evening sky. Flameproof keys were not a myth during that era; they were a necessity!

Signal Electric's Number 2 Key

The trim hand key shown in Figure 4-2 was also made for spark operations. It was manufactured by the Signal Electric Co. of Menominee, Michigan during the early 1900s. It measures $3\frac{1}{4}" \times 2\frac{1}{4}"$ and is solid brass. The key's underside sports an unusual isingglas insulation strip to protect the desk from flames and an authentic Navy knob to protect the operator from slipped (burned?) fingers. The base also has screw holes for mounting (quite necessary during that hearty era!).

Figure 4-3 is a close-up view showing the large(!) silver contacts on Signal Electric's Number 2 spark key. This little gem was truly made for heavy-duty use! If key contacts burned out, or if a light-duty key was pressed into service, dimes were often substituted for contacts!

The Marvelous Marconi 365

The massive Marconi 365 spark key shown in Figure 4-4 truly reflects English craftsmanship at its golden best. This beautiful item is rugged and reliable. Its large contacts handle high current, yet the key feels like a nimble sports car dur-

Figure 4-3. Close-up view of large silver contacts on the Signal Electric. This key was made for hearty use!

Figure 4-4. The marvelous Marconi spark key. This king-size key was used aboard British ships during World War II, but its operators would not let it fade in the annals of time. It was thus used for many years and is a genuine classic. (Photo courtesy K5RW and N4QB.)

ing use. Those are some impressive credentials for such a large key, but our friends across the pond are serious brass pounders! Notice the key's hefty marble base, front nameplate, and solid-brass arm. Notice also that the standard-size knob includes an operator-protecting shroud over the arm's "business end."

This man-size key was mainly used aboard British ships during World War II, but operators vowed not to let it fade in the annals of time. The classic 365 thus continued as a high-seas favorite until the 1960s.

Figure 4-5 is a side view of the Marconi 365. Notice the key's overall size and elaborate ball-bearing assembly at the center. This classic key is being held by its proud owner, Neal McEwen, K5RW. Neal has one of the world's best key collections.

Your First Key (J-38)

The key shown in Figure 4-6 requires miniscule introduction, as it is a popular version of the J-38 many of us used during our early days as radio amateurs. Mounting holes were included on this basic key so it could be screwed down onto a desk or wood block. If you purchased a J-38 from a local army-surplus dealer, it was even premounted on a plastic base. Remember "chasing" that critter around your desk while transmitting? We held keys carefully during that tender age because exposed high voltage across their contacts could rattle your teeth if it was not respected!

The open-frame J-38 shown in Figure 4-6 is mounted on a wood-decoupage plaque for display. It has the circuit-closing lever many of us removed (then sat a Coke bottle on the knob for tune-up!) and

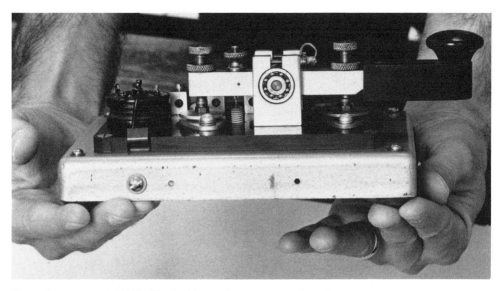

Figure 4-5. Neal, K5RW, holds the Marconi spark key to show its superb ball-bearing assembly and large size. Impressive indeed!

a basic-style knob (remember adding a poker chip under it to homebrew a Navy knob?). The key is all brass (mine even has brass contacts). Thus, a good cleaning with Brasso or Ajax makes it shine like gold, and justifiably so, as the famous J-38 is indeed a golden classic key! Give it a well-deserved fun run during the ARRL's next Straight Key Night. You will love it—again!

What? We are not going to explain how to adjust a J-38? No indeed! In my life I have learned three lessons: Don't pull on Superman's cape, don't tell existing amateurs how to put up a 40 meter dipole, and don't tell a CW buff how to adjust a J-38! Whatever you are doing right now is just fine!

The Pole Changer's Key

The clean-cut key shown in Figure 4-7 is officially known as a Western Union Pole Changer's Key, and it has one unique feature. Notice the lever and triple contacts on the right side. This is not a circuit-closing switch per se, but a contact changeover setup. Move the lever to the right and the key works normally. Move the lever to the left and it keys your rig on "break"

Figure 4-6. Is there an amateur today who does not recognize the famous J-38? Many of us started with these popular hand keys, and they are still found in modern setups worldwide. A book featuring keys would not be complete without a J-38! (Photo via WBØSYV.)

Figure 4-7. This Western Union key was used by field servicemen or pole changers. Note it sports contacts for both "make" and "break," and selection is via a lever adjacent to the knob. (Photo via WB0SYV.)

rather than "make." CW in reverse! That should drive the 40 meter natives bonkers! Who knows? Maybe you could wire it to your rig's FSK generator (up for "mark," down for "space") and send ASCII or RTTY with a hand key! Seriously, however, this is an outstanding key with a great feel.

U.S. Army Key

Another starter key many of today's amateurs remember all too well is the J-5, or closed-case Army key shown in Figure 4-8. These were available at low cost from army-surplus stores nationwide during the 1960s, and original-box leftovers may still be available today.

This key sports a round-top Navy knob and an unusual side arm that works the internal mechanism. Its feel, however, is surprisingly good. The top two screws are removed for adjustments. This key would be a neat match for military-surplus PRC-67 "lunchbox" QRP transceivers. Good luck on finding either or both of these items! The flameproof J-5 in this photo belongs to Charles Tryor, N4LMY.

British Army Key

The spiffy-looking gem shown in Figure 4-9 was rescued from a British army tank by fellow key collector VE7FOU. Rick refurbished it to like-new condition and mounted it on a beautifully finished rosewood base that he inset with lead weights. This key handles as great as it looks, so it sees plenty of on-the-air action today.

Notice the key's classic British knob and use of slot screws with lock nuts rather than conventional knurled-top screws. The key's base is inscribed "KEY WT 8 AMP No. 2 KIII." A magnificent item indeed!

Figure 4-8. Another always-remembered key from our Novice days—the flameproof J-5. Adjustments under top cover. (Photo courtesy N4LMY.)

Figure 4-9. The British army would be proud to reclaim this attractive key which VE7FOU restored to better-than-new condition and uses daily. Check out that spectacular gloss!

Leg-Strapped Army Key

The key shown in Figure 4-10 was used by tank mobile operators in the armed forces and sports a familiar leg clamp for holding it in position during operation. Note the "JJ-37" stamp on its base. We understand some military radiomen also swiveled this key on its mount for walking/portable use while on foot.

Japanese Civilian Key

Precious little information is available on the "generic" Japanese key in Figure 4-11, but it is definitely worth viewing. Notice its heavy arm, precision adjustments, and "hidden wiring."

Hi-Mound HK-1 Key

The professional-looking hand key shown in Figure 4-12 is made by the Hi-Mound Company in Japan. It sports fine craftsmanship throughout, with precision horizontal and vertical adjustments, locking nuts, and a beautiful chrome finish. The key is mounted on a weighted base topped with marble. The Hi-Mound Company is as well known in Japan as Vibroplex is in the United States, so brand-new HK-1 keys may yet be available. They are terrific items to own and/or use, but import taxes can make them quite expensive. For example, I remember an outstanding Swedish key that sold (briefly) in the United States for $120. It faded from the

Figure 4-10. This familiar-looking hand key features a mating leg strap for mobile use. It was used by the U.S. Army during World War II in the Tank Corps. (Photo via JN1GAD.)

Figure 4-11. This impressive Japanese hand key sports their classic "Junior Mint"-looking knob, heavy arm, and precision adjustments. (Photo courtesy JN1GAD.)

U.S. market, but was later available via a Japanese importer for $340.

Figure 4-12's view of the HK-1, incidentally, is courtesy of noted key collector Shigeru Kawasaki, JN1GAD. Shige has a fabulous collection of magnificent keys, and we would like to include more in this book. Unfortunately, while this book was being prepared, several of his photos were at *CQ* magazine's printer being readied for another of our World of Ideas columns featuring classic keys. Want to see more on keys? Read my *CQ* column!

Russian Hand Key

Figure 4-13 shows an impressive item you do not see every day in this country. It is a Russian hand key Dick Randall, K6ARE, purchased for $20 while visiting the USSR in 1969. This key's points of interest include a white plastic Navy knob, black metal arm with center pivot, and rear-mounted tension spring that pulls down rather than pushing up like U.S. keys. Although barely visible in this photo, long black screws with lock nuts secure the arm to the (black) center fulcrum and provide adjustment. The key's base is weighted wood with an attractive finish and holes for screws to mount it on a desk. Although only a rumor, I hear this key transmitted K6ARE's CW in Russian for the first week of use. It settled down and purred beautiful English afterwards.

German Keys

Over the years our German friends have produced some very interesting and popular keys. These items, like German cars, are stout-hearted and built to last a lifetime.

One of their most well-known keys of

Figure 4-12. Hi-Mound's popular HK-1 hand key. This fine-crafted gem is mounted on a marble base, and it works as well as it looks. (Photo via JN1GAD.)

Figure 4-13. Russian hand key purchased by K6ARE during a visit to the Soviet Union. Details in text.

eras past is shown in Figure 4-14. This pint-size marvel is called a mouse key. It was reportedly designed for mobile use in military vehicles and tanks. The key's protective top cover swings up for adjustment. Notice it has a rear pivoting arm, and its travel-setting screw is near the knob. The key's contacts are directly below that adjustment point.

The item shown in Figure 4-15 is an unusual variation of the mouse key. It has a similar concave-top knob, but the arm pivots from the middle rather than the rear. The key snaps into its bottom tray, and then the top cover snaps in place, leaving only the knob exposed.

Two more German keys are shown in Figure 4-16, and both are full-size items for fixed-station use. These keys have center pivots and precision adjustments. Notice the key on the right also sports a domed English-type knob.

Swedish Key

Figure 4-17 shows a superb-feel, precision hand key made in Sweden during the 1950s. The rearmost middle bar is the arm's pivot point (note screw adjustments on the side). The frontmost bar limits and stabilizes arm travel. The rear-arm travel-setting screw has a calibrated knob skirted by a white shroud. A white dust cover fits over the key so that its knob extends out one end and its adjustment protrudes through the top. Very neat!

Figure 4-14. Recognize this little gem? It is the well-known German "mouse key." This key is not only rugged, it also has a terrific feel. Can you visualize using one of these keys in your car? A mobile delight for sure! (Photo via K2EEK.)

Figure 4-15. A unique variation of the mouse key. Both top and bottom covers snap off and reveal a precision item.

Figure 4-16. Two German keys designed for fixed-station use. The left key has a traditional German concave knob. The upper key's dome-shaped knob reflects British influence.

Figure 4-17. An early version of the famous handmade Swedish key. The arm is a single piece separated by two bars. The rear knob adjusts travel. Dust cover protects mechanism.

Figure 4-18. The ideal hand/pump key for knob-twisters! There is only one adjustment on this key from the Netherlands, but it handles fine.

Netherlands Key

What's this in Figure 4-18? A key with no adjustments? Not really. The rear knob sets arm travel. Other adjustments, however, require minor base disassembly. Fortunately, the key has a good feel as is!

Clothespin Key

We all have heard our parents talk about walking barefoot through 3 feet of snow for 8 miles to attend school. The key shown in Figure 4-19 depicts my own "hard-times story."

As a grade-school-age radio amateur, cash money was only a legend. My first rig was thus a one-tuber using parts salvaged from junked radios, bell wire, and popsicle-stick supports mounted on a cigar box. Desperation bred creativity, so I homebrewed a key from a clothespin. How did it perform? About equal to the rig, I would say, but I did wrangle a KP4 QSO before he lost me to Novice QRM on 40 meters.

Home assembly of this little key is a snap, and you can build one or two in a night! Begin by taking one or two of the weakest spring clothespins from mom's laundry bag. Lift one side of the back end and snap apart the two pieces. Drill two aligning holes in each piece, and then drill one more hole in the "top" piece for bolting on a knob. In each hole insert two screws with nuts threaded almost up to their heads. One will adjust travel (and help slightly with tensioning). The other screw will be one (slightly adjustable) contact. Insert a spade lug under that screw. Next mount the clothespin's "bottom half" to a wood block with flat-head wood screws. Add another spade lug for the "bottom" contact/screw. Use a screwdriver to raise the spring, and then insert the key's top and align screw contacts. Finally, for a knob add a wood thread spool cut in half.

This key does not have a professional feel, but it works! If you do not care to use it on the air, mount it on a door and wire it to the doorbell! Make several and use them to decorate windowsills! Use them for (supervised) bird toys! My parrot (a real CW buff) loves them. Note telltale beak marks on the base!

Figure 4-19. The epitome of poverty! K4TWJ's clothespin key snaps together in less than an hour. No claims on feel during use!

Figure 4-20. Notice this key's arm has a hump like a camel. It is thus suitably known as a "camelback key." This item is mounted beside a genuine Bunnell telegraph sounder of the late 1800s. (Photo courtesy K6ARE.)

Camelback Key

This chapter concludes with the classic Camelback key, an old-time telegraphic sounder shown in Figure 4-20. The designation "Camelback" resulted from the key arm's unique hump, a design intended to minimize the "glass arm" syndrome of pre-bug years. Sitting behind a hand key and passing messages for hours on end was not an easy task!

The key and sounder in this photo were made by Jesse H. Bunnell's Company in New York. They are mounted on a dark-wood base and comprise the "business end" of a landline telegraph station. That brings us to the next chapter, which highlights telegraph and code instruments of yesteryear. Some exciting views are featured, so read on!

Summary

These are not all of our highlighted hand/pump/straight keys. More beautiful gems such as special keys, miniatures, and modern ready-for-sale keys are included in the following chapters.

Chapter 5
Telegraph and Code Apparatus of Yesteryear

Most amateurs are aware that Samuel F. B. Morse devised the first telegraphic system, and landline telegraph was a major communications medium for almost 100 years. Unless you are a retired railroad or Western Union telegrapher, however, details of that famous era are usually vague.

This chapter covers telegraphy of yesteryear with brief discussions of Samuel Morse's first setup, overland telegraph systems, Continental/International code differences, and use of Continental code today. It also features a clever converter you can assemble for mating old-time telegraph sounders with modern solid-state transceivers. I am sure you will find this review of telegraphic history fascinating, and it may open yet another exciting door in your key collecting interest.

A second purpose of this chapter is to pass along details of our proud heritage to future-generation amateurs. You can play an important role in this respect by relaying details of our following discussions to young, new amateurs and encouraging them to also relay the tales to next-generation amateurs. Keys, sounders, and stories of past eras will thus be handed down through the annals of time and keep our grass-roots beginnings alive in memories everywhere.

Samuel Morse's First Telegraph

Newer amateurs may visualize Morse code communications beginning with landlines strung across the wild west and electromechanical sounders placed in telegraph offices, but such is not the case. That was after telegraphy was perfected.

Samuel Morse's first telegraph was made using an artist's old canvas stretcher. Wooden clock gears were used to pull a roll of paper across the stretcher while a pencil hanging from a pendulum made zig-zag marks on the paper for dots and dashes. An electromagnet attached to a bar above the stretcher moved the pendulum. The electromagnet in turn was connected to batteries and via wires to the distant telegraph station. The printed-readout concept was used for several years. Then a teenage employee of Morse, James Francis Leonard, discovered the dot/dash sounds could be interpreted by ear. The next evolution thus was replacing the "printer" with the classic sounder.

Morse's original canvas-stretcher system is now on display in one of the largest commercial museums of old-time telegraphy. This museum is located in the mezzanine area of the Western Union Telegraph offices at 655 South Orcas Street, Seattle, Washington. Over 100 items, including early telegraph apparatus, diagrams of systems, photos of stations, wax statues, and early Western Union uniforms, are featured in this unique collection. Be sure to see it if you visit Seattle.

Landline Telegraphy and The Western Union

After Samuel Morse proved the telegraph's viability, numerous small telegraph companies were formed in the eastern United States. The New York and Mississippi Valley Printing Telegraph Company was the largest. It began operation with an impressive 550 miles of wire coverage. Approximately 50 other companies followed suit. There was no interconnection of lines, so messages had to be transferred between companies by runners, and the cost of a single telegram ran as high as $20 (quite a sum in the mid 1800s)! The NYMVPTC thus pursued a unified service to link eastern and western areas. After buying out eleven other companies and their lines, the new name became the Western Union Telegraph Company.

Rapid communications between east (of Missouri) and west (California) became paramount when the Civil War broke out. The famous Pony Express required up to ten days to carry messages across rugged Rocky Mountain terrain and hostile Indian territory. Western Union engineers thus planned and erected telegraph lines across the wild west.

Western Union telegraphers were sharp as tacks. (First they were the pioneers, then they had Ted R. McElroy for inspiration!) Indeed, many operators could recognize each other by the "swing of their fist" and occasional shortcuts/slang in words to get messages across quickly. This outstanding service used telegraphy well into the 20th century before switching over to teletype.

Continental Versus International Morse Code

A frequent comment among newer amateurs listening to the melodious clicking of a telegraph sounder in operation is "That's Morse?" The answer is yes, indeed, but it is American Landline Morse rather than Continental Wireless Morse.

A side-by-side comparison of the two codes is shown in Figure 5-1. Roughly one-third of the letters differ between codes. Why? One has more dits and is more readable by sounders (and more prone to lightning-caused mistakes). The other is more readable by radio. Another interesting point: During the early days of wireless, many operators were required to know both codes and exhibit the ability to switch between them at will.

The two codes also played a noticeable role in the *Titanic* story. Telegraphers aboard the *Titanic* (using Continental code) became irritated at telegraphers aboard a couple of U.S. Navy vessels (using American Landline Morse) and insulted them (remember this occurred in 1912). The Navy telegraphers either switched off their gear or ignored the *Titanic*'s calls, and thus did not respond to their SOS. This fact was uncovered during official investigations after the *Titanic* sank.

Another little-known fact: Prior to the *Titanic* incident, radio/wireless was considered a "frill." Marconi struggled and convinced the *Titanic*'s officers that having a spark rig aboard would be advantageous. Numerous radio amateurs from Maine to Alabama and Florida copied the *Titanic*'s original SOS. Newspapers did not believe their reports, however. The double-hull *Titanic* was unsinkable! After the Titanic sank, officials re-examined wireless/radio's benefits and proclaimed henceforth all ships in excess of 20,000 tons had to be equipped with a wireless communications system. Radio officially came of age!

Telegraph Sounders

The electromechanical sounders employed in landline telegraph systems were (and still are) fascinating gems to see and hear in action. They emit two unique sounds, or clicks: "klick and kalunk." A dot is heard as two klicks close together, and a dash has more time sep-

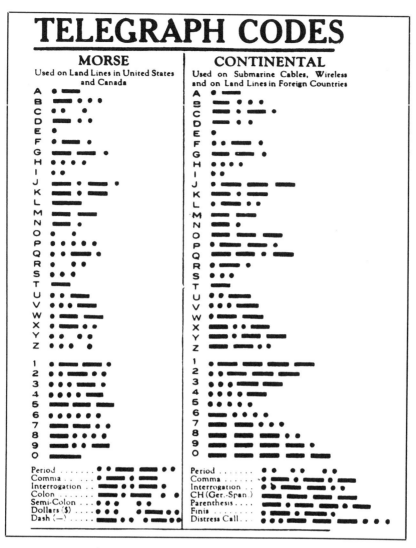

Figure 5-1. Side-by-side comparison of American Landline Morse (left) and International Morse (right) codes. Note differences in various letters and use of l-o-o-o-n-g dashes.

aration with a kalunk at the end. Compare those facts with Figure 5-1. Landline Morse cannot be sent by modern electronic keyers or keyboards. It must be generated by a hand key or bug. Notice, for example, an "L" is a l-o-o-o-n-g dash.

An original-era advertisement of a giant-size Bunnell telegraph sounder is shown in Figure 5-2. These items were manufactured in a wide variety of sizes and styles ranging from "Pony" types to "Longline" models. Some had upright coils, and some had horizontal coils. All are absolute winners. Three to six volts DC usually puts them into operation. The top left screws on our model in Figure 5-2 adjust the "klick/kalunk" sound and arm travel. The right screw (on pedestal) fine tunes arm movement.

Landline Morse sounders are very popular items among key collectors and telegraphers. Tony Isch, W2GDV, found the item pictured in Figure 5-3 at a hamfest fleamarket. The cost was $30, and the sounder works great.

An original Western Union practice set

The Giant Sounder.

J. H. BUNNELL'S
PATENT, FEBRUARY 16th, 1875,
With Improved Adjustments.

We call attention to the fact that our original Giant Sounders, as shown above, have become, since their first introduction in 1875, the standard of excellence throughout the United States, and have also been adopted exclusively upon the Government Telegraphs of England.

They give a loud, clear sound, with just half the amount of local battery generally used on other forms of Sounders.

Price, ... $3 50
Giant Sounders, wound with fine wire to 20 ohms resistance for
Main Line use (without relay), on lines up to 15 miles in length. 4 00

Figure 5-2. Original-era advertisement of Jesse H. Bunnell's Giant Sounder for Landline Morse.

Figure 5-3. Genuine Western Union sounder ready for use. This classic item was spotted at a hamfest fleamarket by proud owner W2GDV.

Figure 5-4. Original Western Union practice set with sounder on left and key on right. The circuit-closing lever on key's right was closed for receiving and opened for sending. (Photo via W2GDV.)

is shown in Figure 5-4. The sounder works just like its big brothers. Generally speaking, there are two varieties of sounders. The "local" types usually exhibit low coil resistance (such as 4 or 5 ohms). They were used on separate circuits in series with a battery and relay on a main circuit. Mainline sounders were larger and exhibited higher coil resistance (such as 30 to 400 ohms). They also included fine-tuning adjustments to compensate for (telegraph) line losses during rainy weather.

A Sounder Adapter for Modern Transceivers

Has landline telegraphy piqued your interest? Great, and you are not alone! Many old-time telegraphers enjoy using American Landline Morse with classic sounders in amateur radio today. Yes, indeed, and you can listen in on the action almost any night. As our friend Art, W8VMX, points out, just tune your rig to 7044 or 3544 kHz for a treat. The only items you will need to add to your setup for copying in authentic old-time style just like other on-frequency stations are a (hamfest-obtained) sounder and an easy-to-assemble converter. The latter's diagram is shown in Figure 5-5. This circuit was passed over to us by Bill Dunbar, AD9E. Bill is president of the Morse Telegraph Club and editor of its quarterly magazine, *Dots and Dashes*. His and the magazine's address is 1101 Maplewood Drive, Normal, Illinois 61761.

This converter connects between the speaker or earphone socket of your super-modern SSB/CW HF transceiver and an old-time telegraph sounder. Parts are available at Radio Shacks nationwide, and their layout is not critical. Transformer T1 matches your rig's low impedance to Q1's high impedance and isolates the two circuits. D1 through D4 assure good operation at low signal levels. R1 is the base bias resistor for Q1. C2 and D5 prevent keying transients. R2 and D6 are a power-on indicator. A 9 volt battery can be used for powering the circuit and sounder, or you can add an external DC supply. An extra resistor can be added in series between Q1's collector and D5's anode for optimizing sounder current, if desired.

Approximate values are 30 ohms for a 70 ma (mainline) sounder, 50 ohms for a 110 ma sounder, and 120 ohms for a 400 ohm coil sounder. Four ohm sounders are not recommended. If you do not have a sounder, a simple 25 ohm/60 ma relay can be substituted to get started (no "klick-kalunk," but you will at least hear authentic American Morse!). If you have additional questions or wish to join the Morse Telegraph Club, write to Bill Dunbar at the above address.

After assembling the converter and giving your sounder a prominent place in the station, tune in the American Morse

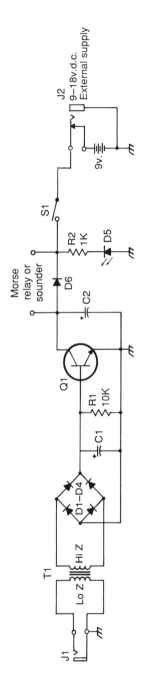

Figure 5-5. A tone-to-DC adapter for mating classic old-time sounders with modern transceivers. Add this adapter to your setup, and it will sound like an authentic telegraph office of yesteryear. See full discussion in text. (Schematic courtesy AD9E.)

PARTS LIST

C1 = 5µF, 35v. (RS272-1012)
C2 = 10µF, 35v. (RS272-1013)
D1–D4 = Bridge rectifier 1a., 50v. PIV (RS276-1161)
D5 = LED (RS276-C66/68/69)
D6 = Silicon rectifier 1a., 600PIV (RS276-1104)
Q1 = NPN, TIP31 (RS272-034)
T1 = Audio output (RS274-1380)

Net on 3544 kHz at 0030 or 1300 GMT, and listen to eras past in high style. Demonstrate it to guests. The impact is amazing!

As this chapter has pointed out, prewireless and landline Morse equipment are extremely popular items among many key collectors. As also discussed, American Morse is alive and well on the amateur bands today. Check it out! You can add some real excitement to your key collecting and amateur radio activity!

Chapter 6
Specialty Items

This chapter highlights some of the more unique and special-interest items that make key collecting and CW operating great fun. Our featured goodies include neat little "spy keys" that fit in the palm of your hand; one-of-a-kind fingerpieces; keys of gold, crystal, and wood; and key-related items such as key fobs, pennants, and coffee mugs. Since these gems do not fit in any official category, I combined them in a potpourri chapter you surely will find appealing.

You might also visualize this chapter as a surprise diversion that always occurs when viewing someone's key collection. You expect to see only straight/hand keys and maybe a paddle or telegraph sounder, but unexpected treats quickly capture your attention and further inspire your own interest in this pursuit. Expanding your horizons always makes life more exhilarating! Now let's "fast forward" to the awaiting views and tales!

Miniature Keys

There is a special magnetism in collecting and using pocket-size "spy keys" that piques the interest of radio amateurs everywhere. Maybe it is the vision of their use with compact battery equipment by secret services or with small spy radios by undercover agents. Maybe it is because palm-size keys are just plain fun to use on the air today (an understatement!). Miniature keys are great companions for portable rigs, and they are custom-made for QRP setups. With proper planning, you can carry a complete QRP station, including batteries and key, in a coat pocket. Now that's traveling in style!

A collection of only miniature keys would be terrific, especially for QRP enthusiasts, and it would be easy to show or carry. Finding these little gems is a real challenge, however. They are usually inexpensive, but hunting grounds are quite elusive.

Precious little information is available on small keys. They seemingly surface right out of the woodwork with very few identifying marks. New owners must thus piece together notes and opinions according to their source and the key's design.

The term "spy key" may also be a misconception. Some pocket keys were originally designed for clandestine use, but some were produced for paratroopers, firefighters, and lifeboat radios. Whatever the case, I am sure you will enjoy viewing the miniature keys in this chapter, and I am also sure you would love using one on the ham bands—if you can find one!

The magnificent little key shown in Figure 6-1 was a gift from fellow collector, Shige, JN1GAD, and I love it! The key measures ¾"W × 2"L. It has a center-pivoting arm with full adjustments, and it feels like a big J-38 during use. On-the-air operations with this key are truly delight-

Figure 6-1. This neat little Japanese "spy key" is the perfect mate for a portable QRP rig. It is only ¾ inch wide by 2 inches long and handles like a J-38! (K4TWJ collection.)

ful. Notice the plastic shroud between the knob and tension-adjusting lockscrew. The shroud covers the arm and adds a touch of elegance. Does the knob's appearance remind you of a big "Junior Mint" like you used to buy at the movies? This style seems to reflect Japanese influence, and although there are absolutely no markings on the key, I am convinced it is a Japanese original. This neat gem has silvered contacts and nickeled upper parts, and the base is heavy plastic. Its workmanship is superb.

The marvelous little key in Figure 6-2 was made by JRC, the Japan Radio Company, and I am confident it is still in production today. The key's proud owner, JN1GAD, sent me a brochure on a new JRC model JST-10 handheld 40/15 meter SSB/CW transceiver, and this key (model CCK-410) was shown as an accessory.

The key's plug is probably "generic," however, and fits other radios.

This JRC key measures only 2"H × 2"W × ¾"D and sports adjustments for both tension and arm travel. It is a rear-pivoting key, but it looks like it has a good feel. The bottom cutout is apparently for clipping on the side or handle of portable radios such as the 12 watt JST-10. This key's knob has an "elongated Junior Mint" shape of Japanese style, and since we know its manufacturer (JRC), we can semi-confirm similar-style knobs (like that in Figure 6-1) are also Japanese products.

Surely the most popular miniature key ever made (and possibly still being produced) is the small Australian lifeboat key shown in Figure 6-3. This item was reportedly packed with battery-operated radios on lifeboats, and no, I have no clue as to

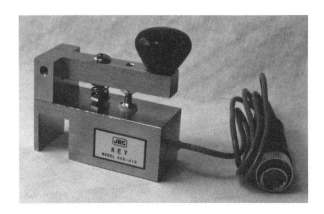

Figure 6-2. JRC's miniature hand key measures approximately 2"H × 2"W × ¾"D and easily slips into a shirt pocket. (Photo courtesy JN1GAD.)

Figure 6-3. This Australian Lifeboat key fits in the smallest pocket. It is completely waterproof and handles exceptionally well. (K4TWJ collection.)

how the two pieces became separated in the past.

This little Australian key measures 1"W × 2½"L, and all working parts are protected by a waterproof enclosure. The knob's shaft even passed through a rubber gasket to the inside, but I removed it to give the key a better feel. All adjustments are precision and accessible by removing the key's top cover and gasket. The inscription on top reads "Transmitting key—light weight—Aust. No. 1." There is also the insignia "A W A" (Australian Wireless Association). Notice the knob's "mushroom" design. This style is quite popular in Australian-made keys. It may look unusual to U.S. amateurs, but it actually is easy to handle. This key works well, and I enjoy using it, but I occasionally have to thump it to stop a continuous string of SOSs. Would I jest?

This key shown in Figure 6-4 was reportedly used in the Indian Telegraph Service, but it has a British-type knob, so I am not sure of its background. The gem measures 1" × 2¼" and has nickel-plated brass parts mounted on a plastic base. It also has contacts for both "make" and "break," indicating telegraph field serviceman use. Arm pivoting is from the rear, but the key handles very well. Its make/break contacts also produce the dual clicks with keying most U.S. amateurs favor. This little key sports a "smokestack"-type knob that is popular in British designs.

The origin and history of the key in Figure 6-5 are unknown, but it is the best-

Figure 6-4. Check out this British and Indian telegraph lineman's service key. It pivots from the rear, has contacts for both "make" and "break," and sports a "smokestack" knob. (K4TWJ collection.)

Figure 6-5. Very little information is available on this ultra-small key, but we understand it was used by railroad detectives in the past. (K4TWJ collection.)

handling miniature I have seen. The key is ½"W × 1¾"L, is all nickel-plated, and even has a circuit-closing side switch. Arm pivoting is near the center, and all adjustments are quite precise. The knob is concave on top, thus reflecting U.S. influence. Wiring connections are made via the long screws. I hold this key upright during use rather than mounting it on a wood block. Why? A base would make the key too large for carrying in a pocket.

These three miniatures shown in Figure 6-6 belong to collector Dick Randall, K6ARE, and they are real delights. The key on the left measures 2" × 2½" and came from a friend of Lee DeForest. The key in the center is only 1 inch square (!) and sports silver contacts mounted in Teflon to reduce noise. It is rumored to be made for the CIA and used with spy-type transmitters. The key on the right is 1½" × 2½". It was used by U.S. forces during World War II. Viewing black keys with black bases is not always easy, so exercise your magnifying glass and good luck finding your own miniature "spy" keys!

Special Fingerpieces

Say you have a favorite hand key, bug, or paddle you would like to dress up in some unique manner? Consider adding custom fingerpieces to the gem, like gunslingers of the old west added special hand grips to their guns.

Lavish, one-of-a-kind fingerpieces are not easy to fabricate, but they are great to own and enjoy. You can invest any amount of money in special fingerpieces,

Figure 6-6. A bonanza of miniatures! The key on the left is 2" × 2½". Middle key is only 1 inch square and is rumored to have been used by the CIA. The key on the right is 1½" × 2½". It was used by secret forces during World War II. (K6ARE collection.)

Figure 6-7. Genuine mother-of-pearl fingerpiece hand-crafted by Rick Van Krugel, VE7FOU. Photos cannot do justice to this remarkable item. All colors of the rainbow glint off it in sunlight. (K4TWJ collection.)

and their originality is limited only by your imagination. The more unique they are, naturally, the higher their cost. Where do you start? Visualize what you would like, and then check with small gem and craft stores that work with the material you prefer. Show them exactly what you want, agree on the price, and then give them time to fabricate the item(s). Understand too that you are taking a chance. An unknown company may take your money and material and run for the border, or the end product may not fit your key. I passed along the idea. You are now on your own for turning it into reality. Maybe the following examples will further spark your creativity.

The Blue Racer bug shown in Figure 6-7 was given a new lease on life with the addition of a genuine mother-of-pearl fingerpiece. Rick Van Krugel, VE7FOU, made the item by laminating pearl strips on each side of a plastic center stock. A groove is cut in the plastic's front so it slips over the bug's lever and secures with the small screw that held the original fingerpiece. Photos cannot do this pearl item justice. All colors of the rainbow glint off it in sunlight or under station lights. The fingerpiece really adds glamour to our home setup. I asked Rick if he would consider making similar fingerpieces for others, and he said yes with hesitancy. Pearl is expensive, numerous man hours are involved in such projects, time is money, and Rick does not have an abundance of either. This fingerpiece is thus cherished at K4TWJ's home.

Figure 6-8 shows a Rick Van Krugel masterpiece! VE7FOU made this pair of fabulous fingerpieces by inlaying hand-cut mother-of-pearl lightning bolts into polished black plexiglas, and they are absolute show stoppers. Countless man hours went into perfecting these twin gems, and they are obviously priceless. The quality workmanship of these items is beyond description. They feel like silk and look like a million dollars. VE7FOU has built and customized musical instruments for many well-known musicians, and his artistry is superb. Oh yes, I use the keyer mobile every week.

The low-budget fingerpieces of Figure 6-9 were used to replace their broken counterparts on an unusual French-made paddle. Do you recognize the materials? They were cut by a plastic name-tag maker at a hamfest. I drew the pattern exactly as I wanted it and according to

Figure 6-8. The custom fingerpieces on this Bencher/MFJ keyer are absolute showstoppers! VE7FOU made them by inlaying hand-cut mother-of-pearl lightning bolts into polished black plexiglas. (K4TWJ collection.)

how their machine cut, gave the vendor the pattern one Saturday morning, and carried the fingerpieces home that afternoon. Although they work fine, the finished products would be more impressive if they were glued to a heavy-plastic center section. Jade or synthetic-ivory fingerpieces should also prove unique. Think creatively, and fingerpieces of your own design will be real conversation starters.

Homebrew Electronic Paddle

Going a step beyond fingerpieces, VE7FOU homebrewed the complete electronic paddle shown in Figure 6-10 right from scratch. The end product works like a dream. Rick made the paddle using flexible metal strips as arms. They are spaced and insulated by a center plastic support. Screws and nuts holding the fingerpieces contact the center/common ground post—simple and effective.

Figure 6-9. An example of custom fingerpieces made from readily-available material is shown on this French "El Key." They were cut from plastic nametag stock by a hamfest vendor.

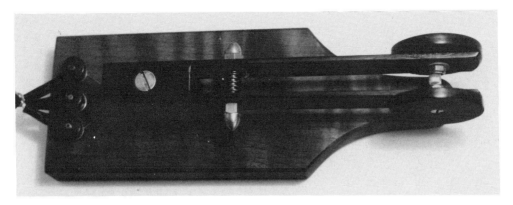

Figure 6-10. VE7FOU's completely homebrew paddle for use with an electronic keyer. Item was made from rosewood and metal strips. (VE7FOU collection.)

The paddle is mounted on a beautiful rosewood base weighted with lead inserts.

Gold-Plated Key

The item shown in Figure 6-11 is hand made in England by well-known key craftsman Gordon Crowhurst, G4ZPY. It is a beautiful gold "trophy model" of his popular hand/pump key described in the next chapter. This key is perfect to the finest detail, including its ball-race pivot point, and it is a real working model. The key is enclosed in a glass trophy case that is also available with this item. Gordon makes these keys on special request and according to available time. They make a magnificent addition to any amateur's setup.

Crystal Key

Figure 6-12 shows a very unusual item you do not see every day. It is a cut-in-crystal version of a hand/pump key sitting atop a half-dome symbolizing the world. This unique paperweight was made in Russia and presented to Evelyn Garrison, WS7A, of ICOM America by the Soviet competitors in The World Radio Team Championship 1990. Inscribed inside the dome is "AREL," the USSR equivalent to

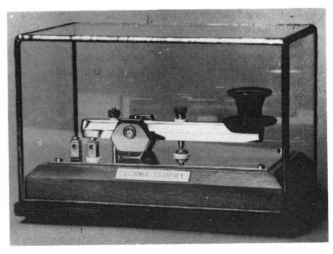

Figure 6-11. Solid-gold "trophy" model key hand-made by G4ZPY of England. This working item is enclosed in a glass case. Photos cannot capture the full glamour of this gem. It is beautiful!

Figure 6-12. This cut-in-crystal version of a hand key sitting atop the world was made in Russia. It was presented to Evelyn Garrison, WS7A, by the Soviet competitors in the World Radio Team Championships in 1990. (Photo courtesy WS7A.)

our own ARRL. This attractive display item may well be the only one of its kind in existence.

Wild Woody WARC Keys

My own wacky contribution to the wonderful world of CW is shown in Figure 6-13. While writing this book I had several hundred of these Wild Woody WARC Keys custom made and "retained the molds" to make more when the supply gets low. Why? Just for promoting light-hearted amateur radio fun and enjoyment! They are fashioned after my first ham key (described in a previous chapter), but times sure have changed. My original wooden keys could be assembled for less than a dime. This 1990 equivalent—its knob, base, WARC insignia, and mailing box—tallied almost $1.00 each!

I plan to award these Woody WARC Keys to amateurs working us on two WARC bands or two HF bands and one WARC band during my Wild WARC Weekends or WARC Days of 1991 and future years. I will also give out WARC keys during my upcoming special events operations, IOTA expeditions during stand-up CW contests at hamfests, and at any other legitimate-excuse time. Ask about a

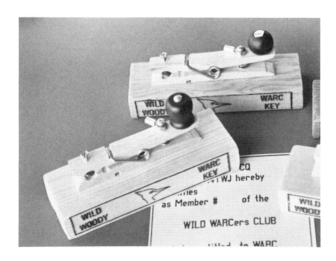

Figure 6-13. K4TWJ's Wild Woody WARC keys! Your author will be awarding these wacky warblers to deserving amateurs in the future. (Details in text.)

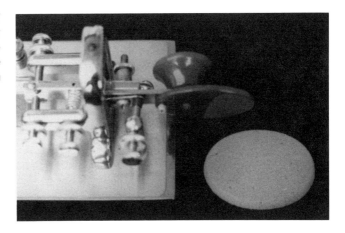

Figure 6-14. The ultimate CW accessory—a windscreen for your key's fingerpiece! Maybe we should offer this unique item for sale on April 1 of each year.

Wild Woody WARC key when we work each other on the air or meet at a convention. Then you too can go WARCing along with your own Wild Woody WARC Key!

Windscreens for Bugs and Paddles

If you pursue big-time CW mobiling or outdoor operating, the item shown in Figure 6-14 is your ultimate accessory. This foam windscreen slips over your key's fingerpiece and works just like its equivalent for microphones to reduce wind noise or road rumble on CW. A secondary benefit is it gives your fingers something soft to handle while transmitting.

This sweet little nothing can be homefabricated from thin foam packing if you are artistic with scissors. Alternately, commercial counterparts have been spotted hiding in ladies cosmetic counters of larger stores. We will leave the excuse for purchasing the latter to your creative imagination.

Vibroplex Goodies

Vibroplex has an impressive array of key-supporting accessories guaranteed to

Figure 6-15. Two popular Vibroplex goodies you can purchase now: a 22-karat-logo mug and solid-brass key fob.

enhance your love of CW. Two of their most popular items are shown in Figure 6-15. The mug is black glazed stoneware emblazoned in 22-karat gold with the Vibroplex logo. The oval is a solid-brass key fob with the famous Vibroplex logo.

Additional items not shown include a Vibroplex pennant (it looks great over your key collection), a solid-brass belt buckle, Tee-shirt, and cap. Vibroplex also sells plastic dust covers and carrying cases for their keys. Check with them for more information.

Summary

This chapter's diverse look at special treats associated with key collecting and CW operating covered quite a bit of ground. I hope you enjoyed the views, and I trust it inspired your creative imagination in adding some unique touches to your own setup.

Chapter 7
Modern Keys and Paddles

We now focus on the easiest, fastest, and most convenient way to begin or improve your key collection—purchasing new items directly from their manufacturer or dealer. This market is larger than many amateurs realize, and some real treats are available to suit everyone's taste. Their outstanding quality and combination of brass, wood, chrome, and marble make a spectacular display. The keys are also tremendous fun to use on the air, and create the same exhilaration as driving a new car. These items are not low in cost, but you can add one a year to your collection or purchase only one or two keys for greater on-the-air enjoyment.

Are new keys suitable for collections? Yes, indeed! There is absolutely nothing improper about a collection of even all new keys. In fact, it makes good sense to enjoy using the latest and greatest goodies right along with your favorite oldie. As we also discussed, key collections reflect you and your interests. We know several amateurs who collect only electronic paddles, and some who collect only new hand keys. If a special area of key collecting captures your interest, go for it and enjoy the results!

Proceeding with this philosophy foremost in our minds, let's look at some beautiful modern-day delights destined to become genuine classics. I am sure there are more keys than just the ones we show in this chapter, especially imported keys. Please forgive us for overlooked items. If you will share those views with us, we will include them in a future edition. Now let's shift into high gear and move on to the photos!

Vibroplex Paddles and Keys

The first name in semi-automatic keys, Vibroplex, continues proud and strong today. The company's line has changed with the times, but Vibroplex is still a world leader in precision CW instruments. As pointed out in Chapter 1, the "Original" model bug invented by Horace G. Martin in 1890 is available from Vibroplex dealers worldwide. This classic key is made in three versions: the "Standard," which sports chrome upper parts mounted on a gray base, the all-chrome "Deluxe" with jewel movement and red knobs, and the gold-based "Presentation." The latter has everything—gold, chrome, jewel movement, and an adjustable mainspring. These beautiful keys are made using the same tools and dyes as Martin's first keys, they handle like silk, and they are the true symbol of a CW expert.

Continuing the Original's famous design concept, Vibroplex also manufactures two similar-looking styles of paddles for electronic keyers (see Figure 7-1). Their "Vibrokeyer" is a non-iambic and single-lever paddle closely resembling the Original model bug in appearance. This is an ideal first paddle for new operators or old-timers accustomed to

Figure 7-1. Vibroplex's beautiful single-lever paddle is truly a modern-day classic. This elite item is readily available today in Standard, Deluxe, and gold-based/Presentation versions.

semi-automatic keys. The Vibrokeyer is available in a Standard/gray base or Deluxe/all-chrome version.

Vibroplex also manufactures a dual-lever paddle designated the "Iambic." This item has a great feel, and it is the perfect mate for electronic keyers with automatic dot/dash insertion and memory. You simply tap the proper fingerpieces, and the setup sends perfect code regardless of "shaky fingers." The superb quality of this paddle is truly legendary, and many operators consider it the Cadillac of the industry. Vibroplex's Iambic model paddle is available in three versions: the Standard, Deluxe, and gold-based Presentation.

Two additional items were introduced into the Vibroplex line in recent years: the Brass Racer Iambic and the Brass Racer EK-1 (Figure 7-2). The former is a smooth-handling paddle for use with an external electronic keyer. The latter is a completely self-contained iambic paddle and keyer. It uses a popular Curtis 8044 keyer IC, and positive or negative keying polarity is selectable. A small knurled knob on the side sets keying speed. Both model Racers feature magnets instead of springs for setting tension. For more information, contact Vibroplex, 98 Elm Street, Portland, Maine 04101.

Bencher Paddles

Another U.S.-manufactured paddle of outstanding quality is the well-known Bencher model shown in Figure 7-3. This dual-lever iambic paddle features adjustable tensions and spacing, gold-plated contacts, and self-adjusting needle bearings. The upper parts are solid brass with heavy chrome plating. The base is heavy steel. Notice the unique split ring with attached contact arms—a space-saving design that also ensures top performance.

The popularity of Bencher paddles continues to increase daily, and with good reason. They are fabulous items with the responsiveness of a nimble sports car! The iambic Bencher is available in a standard/black base with chrome upper parts version (BY-1), all-chrome version (BY-2), and all-gold version (BY-3). A non-iambic Bencher paddle (same design, but no split ring) is available in an all-chrome (ST-2) or all-

Figure 7-2. The Vibroplex Brass Racer is presently available in two versions: an iambic paddle and a completely self-contained electronic keyer. Both versions use magnets in lieu of springs for tensioning.

gold (ST-3) version. For more details on these modern classics, contact Bencher Inc., 333 West Lake Street, Chicago, Illinois 60606.

MFJ/Bencher Combination

Expanding on and complementing an already popular item, MFJ Enterprises presently manufactures a model 422 keyer that fits onto a Bencher paddle. The resultant product is a stand-alone keyer/paddle with Curtis 8044 IC, dot/dash insertion and memory, built-in speaker, and 9 volt battery. This best-of-both-worlds item is available in two versions: the (complete and ready-to-use) 422 with Standard Bencher, and the 422X keyer-only that slips onto your existing Bencher paddle. Either way you get a unit that is great for home, mobile, or portable use.

MFJ presented six of their model 422 keyer/paddle combinations to the winners of the World Radio Team Championship competition held in Seattle, Washington in 1990. The keyers carried WRTC's special logo and were labeled "Winners." One of those six keyers is shown in Figure 7-4. A couple of months later MFJ changed the silk-screen pattern of those 422 awards to "Commemorative Edition" rather than "Winners" and offered them for sale at 422 prices. A limited number of these WRTC-logo keyers were made.

For more information on MFJ's line, contact them at P.O. Box 494, Mississippi

Figure 7-3. Bencher's ever-popular paddle looks and handles great. It is available in both iambic and non-iambic versions with standard/black, deluxe/chrome, or ultra-deluxe gold base.

Figure 7-4. Special WRTC '90 version of the popular Bencher/MFJ combo keyer. Commemorative editions of this item were made by MFJ Enterprises.

State, Mississippi 39762 (phone 1-800-647-1800).

G4ZPY Keys and Paddles

A small but very distinguished line of pump keys and paddles is presently being hand made in England by Gordon Crowhurst, G4ZPY, and they are definitely modern classics. Gordon's quest for perfection is reflected in every one of these keys. They are precision items with fine adjustments, look like expensive jewelry, and handle like a dream.

G4ZPY's most popular item is the hand/pump key shown in Figure 7-5. The key's arm and pivot assembly sport a unique octagon shape. Permanently lubricated bush bearings are included at the pivot point. The black top-hat knob is complemented by a clear plastic skirt. Several options are available when you purchase this masterpiece. You can order the key with diamond-polished brass upper parts or with chrome, silver, or gold plating. You can also choose a Lakestone (green), marble, or mahogany wood base. The quality workmanship of this key is fantastic.

Another very impressive item is G4ZPY's single-lever (non-iambic) paddle shown in Figure 7-6. This little gem sports a unique spring mounted on the rear of its main arm for tensioning. The arm rides on tiny ball bearings set into its support yoke to give a great feel during use.

Although adjustable, dot/dash contacts are preset at one to three thousandths of an inch (with automotive feeler gauges, no less!) for feather-touch action. The paddle thus is perfect for medium- to fast-speed Morse operation. This paddle is also available in brass, silver, chrome, or gold. Its standard steel base

Figure 7-5. This attractive and smooth-handling pump key is custom-made by G4ZPY. It is available in several styles and options as discussed in the text.

Figure 7-6. A unique single-lever/non-iambic paddle made by G4ZPY. Notice the spring fitting against arm's end for tensioning. The arm also rides on a tiny ball bearing at the yoke/pivot to produce a great feel.

is painted glossy black, and the fingerpiece is white plastic. Overall, the item is extremely attractive.

The G4ZPY twin-lever paddle shown in Figure 7-7 handles even better than it looks. This little marvel has a split rear disc for iambic operation and separate dot/dash spring adjustments for optimum tensioning. Concave-to-convex surfaces at pivot points produce a smooth rolling action and wear in rather than out with use. This paddle is available with polished brass, silver, chrome, or gold-plated upper parts, with a matching base or a standard glossy black base. A three-wire minijack under the paddle makes hookup quick and easy.

G4ZPY also makes a very high-speed twin-lever paddle CW buffs will love. This magnificent instrument is shown in Figure 7-8. It sports non-flexible paddle arms fitted with extra-thick and wide-spaced oval fingerpieces. Contacts are silver-to-silver with absolutely no backlash. Although this feather-touch paddle works well at slow speeds, it really comes alive at speeds above 40 wpm. You can order this paddle in chrome, silver, or gold.

Additional information is available directly from Gordon Crowhurst, G4ZPY, 41 Mill Dam Lane, Burscough, Ormskirk, Lancs, L407TG, England (telephone 0704-894299). G4ZPY key brochures and price lists are also being distributed in the United States by assistant Charles Tryor, N4LMY, 7809 Tenth Ave. South, Birmingham, Alabama 35206. Write to Charles for a quick reply to any questions about the keys, and then order a key from Gordon. (Hint: Use your Visa or MasterCard and let them make the dollar-to-pound exchange. It is the easiest and least expensive method of purchase.) Gordon is very exacting in his work. You may face a month's delay in receiving a key, but it will be worth the wait.

Schurr Keys and Paddles

A superb and very dedicated German machinist named Schurr makes some outstanding keys in his basement, and they are available from his representative, Klaus, DL7NS. I have used two of these hand-made Schurr keys for several years, and they are absolutely fabulous. Their quality and workmanship are incomparable. Mr. Schurr gets a tad behind

Figure 7-7. G4ZPY's twin-lever paddle is an iambic delight. It may look like a Gatling gun from the front, but it handles great.

Figure 7-8. The G4ZPY high-speed twin-lever paddle. This key has a feather touch, no backlash, and really comes alive at speeds above 40 wpm.

filling orders at times, but his keys are worth the wait.

Schurr's neat little Mobil hand key is shown in Figure 7-9. This beautiful item is all brass, diamond-polished to the quality of pure gold, and coated with "zaponierung" to retain its luster. The concave knob is cherrywood, the contacts are sil-

Figure 7-9. Schurr's hand-made Mobil key is solid brass diamond-polished to the luster of pure gold. This pocket-size key handles great.

Figure 7-10. Schurr's new Champion model hand key. Sealed ball bearings are set into the fulcrum, and the key's workmanship is superb.

ver, and adjusting screws are very precise. There is absolutely no backlash in this rear-pivoting key, and many operators say this hand key has the best feel they have ever experienced. The Mobil key measures 1"H × 1⅜"W × 3½"D, and comes with a 3 foot flexible cable ready for soldering to your plug.

Schurr's recently introduced Champion hand key is shown in Figure 7-10, and this full-size item is also a dream. All working parts are diamond-polished brass. The base and knob are cherrywood. The center fulcrum's pivot assembly includes permanently lubricated and sealed ball bearings for ultra-smooth operation.

Schurr's dual-lever paddle, or Wabbler, as it is called in Germany, is shown in Figure 7-11. I cannot suppress my enthusiasm for this delight. It's great! I have an early model (painted metal base) and love it. The new model Wabbler is solid diamond-polished brass. The paddle's dual levers pivot on hardened-steel needles force-fit into their molded vertical tubes. The tubes in turn are force-fit into the paddle's base and triangular yoke. Backlash is nonexistent. Keying is positive and effortless. A single spring po-

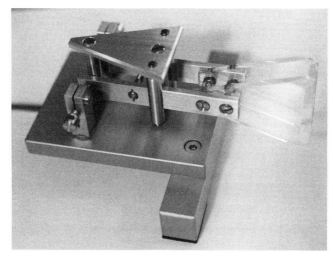

Figure 7-11. This early dual-lever Schurr paddle features diamond-polished brass parts mounted on a painted gray base. It has been used and abused at K4TWJ's QTH for several years, but still handles like new. Newer versions are all-brass and built to last a lifetime.

sitioned between the two arms and fitted with an adjusting nut controls tension. The little paddle is heavily weighted, and does not walk with the most stringent use. A plastic shroud is also available as an option.

For more information about Schurr keys and paddles, write to Klaus Gramowski, DL7NS, Kaiserin-Augusta-Allee 91, D-1000, Berlin 10, Germany.

Kent Keys and Paddles

Kent Ltd. of England is well-known to CW enthusiasts worldwide. They make an impressive man-size pump key and two styles of paddles, and to the best of my knowledge, Kent is the only company today offering their goodies in kit form. Don't cringe. Kent's classics are not like the old Electric Specialty Company's Speed Bug kit. All the hard parts (pivots and bearing assemblies) are preassembled. All you do is mount the main parts in predrilled holes. If I can do it, so can you! Alternately, factory-assembled Kents are available for a few dollars more.

My own (assembled) Kent pump key is shown in Figure 7-12. Its solid-brass mechanism includes a ball-race assembly and large silver contacts. Every piece of this key, including its fine-thread adjustment screws, is made to exacting specifications. It exhibits beautiful workmanship and loud "click-click" sounds on "make" and "break." During use, it even reminds me of an old-time telegraph sounder (not a key for sneaky night use!). The key's mechanism is mounted on a high-luster and lead-weighted wood base. A felt-backed metal bottom lets the key sit flush on a desk, or you can raise it with (included) rubber feet. Although Kent's key is large (3"H × 3"W × 8"D), it has a very good feel. A view of my Kent key prior to assembly is shown in Figure 7-13 to inspire your interest. Assembly time was roughly 30 minutes.

Kent also makes a single-lever and a dual-lever/iambic paddle (Figure 7-14). These paddles, like the pump key, are quite rugged and handle great. For more information, contact Kent Ltd., Tarleton, Preston, PR46YB, United Kingdom.

The Legendary Swedish Key

Precious little information is available on the hand-made Swedish key in Figure 7-15. Although I do not have a valid address for ordering this gem, its photo definitely warrants sharing. This item is the

Figure 7-12. Many of you surely recognize this popular Kent pump key. It is quite large and loud, but terrific fun to use on the air.

Figure 7-13. My Kent pump key prior to assembly (discussion in text).

unquestioned master of hand keys. It is built better than a Swiss watch and includes a unique rear tension spring with knurled knob for adjustment. I had an opportunity to use one of these marvels a few years ago, and I have yet to find another hand key of comparable quality. It was comfortable, smooth, and very nimble. I could even send bug-type Morse at high speed with the key!

This item was imported into the United States for a brief period, but then it disappeared from the market. It later appeared on the Japanese amateur market. At one time a U.S. importer was rumored to be bringing the key back into the States—via Japan! Good luck finding one of these golden classics!

Figure 7-14. Kent's single- and dual-lever paddles are rugged and handle fine. Their ball-bearing assemblies are similar to those used on Kent pump keys.

Hi-Mound Keys

Your super-sleuth tracking abilities will again be required for locating a viable source for the remaining keys highlighted in this chapter. Regardless of your success, however, I am sure you will appreci-

Figure 7-15. This legendary Swedish key is almost as famous as the Swiss army knife, but it is rapidly fading from the U.S. market.

ate reading about these modern-day classics from the Orient.

Hi-Mound is a very popular and respected name in transmitting keys in Japan. This company has been in business for many years and is as well-known in the Far East as Vibroplex is in the United States. Hi-Mound produced a couple of semi-automatic keys in eras past, and at least one model (highlighted in Chapter 2) sported their early Swallow nametag. Hi-Mound concentrates on hand/pump keys and electronic paddles today, and their products are absolutely top of the line. Maybe you can purchase one during your next visit to Japan. Otherwise, enjoy our views of "DX Keys"!

Our first two Hi-Mound keys are shown in Figure 7-16. The lower item is a "palm keyer" especially designed for in-hand use while operating portable or mobile. Although its exposed lever looks like it actuates an internal microswitch, owner JN1GAD assures me a full and completely adjustable hand-key mechanism is in the case. The upper item is an early Swallow nametagged pump key. Outstanding quality and superb craftsmanship are immediately apparent in this gem. Its "Junior Mint"-type knob definitely reflects Japanese style. I understand this key handles great.

Another pump key with the Swallow nametag is shown in Figure 7-17. This neat little key sports full adjustments on all moving parts and a beautiful marble base. The key would add 3 dB of glamour to any setup, even if it was not used on the air!

A later-model Hi-Mound pump key is shown in Figure 7-18. This item features very precision adjustments and terrific chrome plating. In addition to its heavily weighted base, there are four countersunk holes for screwing it to a desk.

A miniscule amount of information is

Figure 7-16. Two outstanding Hi-Mound hand keys available in Japan. Palm key (lower item) has a complete mechanism inside the case. Upper key carries classic Swallow nametag. (Photo via JN1GAD.)

Figure 7-17. Chrome parts mounted on a marble base make the Swallow pump key glamorous and delightful to use. (Photo via JN1GAD.)

available on the two pump keys shown in Figure 7-19, but they are too attractive to overlook. These keys are part of JN1GAD's collection, and judging by the split arm, I would say they are Korean-manufactured items. Such learn-as-you-go observations are pointed out here because they accurately depict key-collecting techniques. If you acquire one new particle of useful knowledge from any key viewed, you are doing well.

The Kenpro Key

The impressive chrome and marble item shown in Figure 7-20 is also made in Japan. It is manufactured by Kenpro, a familiar name in imported antenna rotors. Notice the key's "Junior Mint"-type knob with navy skirt. The key's mechanism is protected by a round plastic cover that snaps off for adjustment. We understand this gem handles exceptionally well.

Summary

We trust you enjoyed viewing this chapter's array of modern-day classic keys and paddles. Some are easy to obtain, while some are not. Such is the game of key collecting!

Figure 7-18. Late-model Hi-Mound pump key with exquisite chrome plating and precision adjustments. (Photo courtesy JN1GAD.)

Figure 7-19. Korean-manufactured pump keys. Note unusual split in arm's rear for holding the travel adjusting screw. (Photo via JN1GAD.)

Figure 7-20. This Kenpro key has a very good feel, and its clear dust cover snaps off for adjustments. (Photo via JN1GAD.)

Figure 7-21. This is the phenomenal Mercury paddle. These beautiful items are hand-made by Steve Nurkiewicz, N2DAN/4, (1385 Abner Street, Port Charlotte, Florida 33980). They have magnets instead of springs and are works of art. (Due to the late arrival of this information, there are no details in the text. Contact Steve for more information.)

Chapter 8
Classic Rigs For Your Classic Keys

Whether your collection of classic keys is large or small, sparkling clean or a mite dusty, nothing compares to using those beautiful oldies with similar-era rigs on the air today. There is a special magic to units with soft glowing filaments and dim amber dials that cannot be equalled by deluxe solid-state transceivers with cold transistors and digital readouts. There is no doubt modern transceivers are more effective and easier to use. That is unquestioned, understood, and accepted. Enjoying an occasional operating diversion with an old-time setup, however, adds new excitement to your amateur radio world.

Do not just take my word for these statements. Try it yourself and see! Experience the thrill of sparking key contacts, winking antennas, and howling earphone-receivers. Learn how to copy one signal amidst many by listening to a single tone or swing of the other operator's fist. Feel the exhilaration of working DX with only 5 or 10 watts output. That is amateur radio in high style, and it truly proves the operator rather than the rig makes the difference!

Where do you begin in setting up such a classic amateur station today? That depends on your own nostalgic interest, age, and technical expertise. If you really appreciate old-time radio at its golden best, a 1930s-style rig is fabulous fun. It is also cantankerous to get going, and assistance from related-era magazines or old-timers familiar with self-excited gear is heartily encouraged. Early-era homebrew receivers typically "miss" their targeted range, and experimenting with coil turns and capacitor values is vital. Transmitters also lack bandspread and cover several megaHertz without respect for amateur band edges.

These vintage transmitters and receivers are thrilling and challenging to operate. If you feel uncomfortable using such gear, however, don't. Go with your instincts and switch to a crystal-controlled transmitter plus a more predictable 1940s/1950s-vintage receiver.

Rebuilding your original Novice rig is another great idea. Maybe you did not work much DX with it when you started out, but you can now apply expertise acquired over the years and have a ball making long-range contacts. A little 10 watt rig in the right hands and connected to a good antenna can work the world!

Bearing in mind the previous thoughts, let's briefly review a 1930s-style station you can duplicate today. Following that discussion, we will highlight an ever-popular 6L6 transmitter you can mate with a 1940s/1950s-style commercial receiver. If you need additional construction guidance or become hopelessly addicted to tinkering with old-time amateur gear, please refer to original-era handbooks and magazines. These items are often available at local-area libraries.

Another lighthearted review of beautiful old-time equipment is featured in my recent book *Golden Classics of Yesteryear.* Copies are available from MFJ Enterprises (1-800-647-1800) and from dealers nationwide. Now return with us to those thrilling days of yesteryear when untold radio thrills lurked behind every turn of the dial, and every night's operation was an exciting new adventure. Old-time radio with all its golden glamour lives again!

A 1930s-Style Hartley Transmitter

The following transmitter was well-known for its outstanding DX performance during the early 1930s. Thanks to its rigid construction and hefty plate coil, it still puts out a respectable CW signal on today's bands. The transmitter's schematic diagram is shown in Figure 8-1, and a photo of this classic rig beside its mating 30-30 receiver is shown in Figure 8-2. The transmitter's two tuning condensers are mounted vertically rather than horizontally, and the tube socket is mounted on "L" brackets bolted to their frame. Filament bypass and plate/grid acorn condensers are mounted below and behind the tube socket with metal straps, and the plate choke is positioned vertically beside the left/main-tuning condenser. Glass curtain rods and their wood mounts, popular five-and-dime-store items of the 1930s, support the copper plate and antenna coils to prevent vibration and assure a stable signal. An RF ampmeter is strapped to, but insulated from, the right/antenna-loading condenser for indicating output power (the old I^2R method).

The plate coil consists of ⅛ or ¼ inch copper tubing wound with an inside diameter of 2¼ inches and spaced or stretched to a length of 6 inches for 80 meters or 4 inches for 40 or 30 meters. A 12-turn coil and 400 mmFd (that's pFd for newer amateurs) condenser will cover 80 meters, 6 turns and 200 mmFd will cover 40 meters, and 5 turns plus 150 mmFd will cover 30 meters. The antenna coil is 5 turns

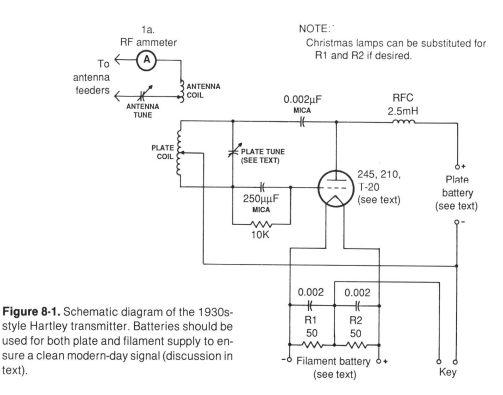

Figure 8-1. Schematic diagram of the 1930s-style Hartley transmitter. Batteries should be used for both plate and filament supply to ensure a clean modern-day signal (discussion in text).

Figure 8-2. The 1930s-style Hartley transmitter beside its mating 30-30 regenerative receiver. Note Mac-Key and Baldwin earphones. This is going in style, and it really proves the operator rather than the rig makes the difference!

for 80 and 40 meters, and 3 turns for 30 meters. It is also copper tubing like the plate coil. Operation above 30 meters is not suggested with this rig, as stability and in-band operation are unpredictable. One spin of the dial can whisk a signal from 12 to 16 megacycles!

Since circulating RF tank current can approach several amperes in this little rig, the plate coil and its tuning condenser must be bolted together firmly to minimize heat losses. Large bolts and wing nuts were used in the original version, but I fudged simply by flattening and drilling coil ends and using the condenser's existing bolts for mounting. Band changing is more difficult, but I only use this little rig on 30 meters.

My home re-creation of this beauty is fairly close to the original, but some sacrifices were unavoidable. The tuning condensers are *big* Hammarlund items rather than the very scarce Cardwells, the glass curtain rods are faked with plastic dowels, and I am still searching for a 2 inch RF ampmeter.

Several types of triode tubes such as the 210, 201, or T-20 work fine in this transmitter, but I prefer using an ever-popular type 245. This little 5 watt tube is usually easy to find, and it works well with plate voltages between 100 and 300 volts. A bank of fifteen or twenty 9 volt batteries series-snapped together makes a good "B supply," and a 3 volt dry cell with a series-added 2.5 ohm 10 watt resistor will power the 245's 2.5 volt filament. If you prefer an AC supply, use a transformer with at least 100 ma current rating plus 20 to 40 mFd of filter capacitance to ensure good dynamic stability (no chirp). Also, avoid powering 245 filaments directly from an AC source, as it promotes hum. Rectify and filter it!

If you would like to add some extra sparkle to this rig (not necessarily for on-the-air use!), consider homebrewing your own electrolytic rectifiers for the AC supply. Now this is big-time radio in high style! Home-grown rectifiers were described to me by genuine old-timer W5OC as follows.

"First mix 20 Mule Team Borax and water into a semi-liquid solution, and pour it into some of mom's fruit jars. Insert a 1 or 2 inch aluminum plate on one side of the jar for a positive terminal, and add a lead plate on the opposite side for a negative terminal. Connect the rectifier as usual between your transformer and filter condensers. Each rectifier is good for 50 volts. Use four series-wired jars for 200 PIV (or 16 jars if you build a full-wave bridge). Baking soda is a fair substitute if you cannot find Borax (a famous sponsor of TV's early *Death Valley Days*)."

These fruit-jar rectifiers emit a fantastic bluish glow that resembles an outdoor bug zapper following your keying. Combine that with arcing key contacts and a wild-swinging iron vane meter, and you have a real 1930s-style rig visitors will love (or fear!). Add a small NE-2 neon lamp to this rig's mated dipole antenna, and then look out your window at night and see your signals headed for distant lands.

A Mating 30-30 Receiver

The classsic receiver shown on the right in Figure 8-2 uses a pair of famous old 30 tubes, and it works like a champ. The original model 30-30 was built "open-air" style on a 10 inch square breadboard with its tuning dials mounted via angle brackets. You can update that design to mid-1930s standards with an aluminum chassis and front panel.

Now refer to the schematic in Figure 8-3. The AF transformer was any on-hand item with a 1:3 to 1:6 turns ratio. Some folks used transformers salvaged from old radios, while others used doorbell transformers. A most forgiving circuit indeed!

Condenser C1 is the antenna trimmer, C2 is the main tuning control, and C3 is the regeneration control. The RFC is a regular 1 or 2.5 mHy item. A medium-size 10 ohm rheostat, R1, is wired in series with the filament battery to drop 3 volts down to 2 volts for the number 30 tubes. Two 22½ volt batteries are used for the "B" supply. A tap "between" the two batteries provides 22½ volts for the detector, while both batteries deliver 45 volts to the amplifier. Since the amplifier tube's

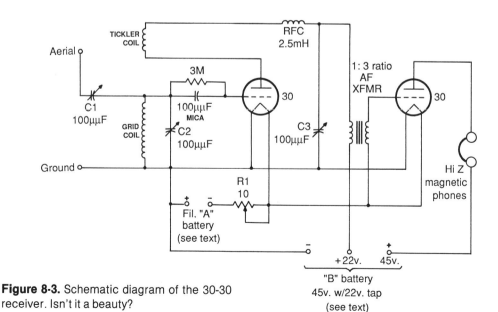

Figure 8-3. Schematic diagram of the 30-30 receiver. Isn't it a beauty?

Figure 8-4. Coil data for the 30-30 receiver.

BAND	NUMBER OF TURNS	
Meters	Grid coil	Tickler coil
100–200	54	20
50–100	24	13
25–50	12	8
15–25	6	7

plate load is the earphones, old-time high-impedance types such as the classic Baldwin "cans" must be used. You can experiment with a small 100K to 8 ohm transformer with modern earphones, but volume loss may prove excessive. Check with Antique Electronic Supply Company in Tempe, Arizona or search hamfest fleamarkets for high-impedance earphones and other parts from yesteryear.

All coils for this neat little receiver are wound on standard four-prong forms 1¼ inches in diameter and 2⅛ inches in length (see Figure 8-4). Homebrew forms can be made from tape-fattened tissue rollers and old tube bases. Paint them brown for authenticity. Number 24 double-cotton-covered wire was used in the original coils, but modern substitutions such as number 20 or 22 enamel-covered copper wire may be required today. Space out the turns along the form to assure maximum bandspread, and wind the grid and tickler coils in the same direction on the form. Separate them ⅛ inch apart, with tickler near the bottom.

Your exact tuning range may vary slightly from original-era coils, but finding the band(s) is a natural part of building and using classic rigs! Simply advance the (completed) receiver's regeneration control until the earphones howl, and then use a little Digitrex frequency counter or your modern rig's general-coverage receiver to check its radiated signal. (Regenerative circuits emit a weak signal on their receiving frequency. That signal can be used to determine its frequency range and calibrate its tuning dial.)

Lay out the receiver with the antenna input on the left, main tuning condenser in the middle, and amplifier stage on the right. If the set fails to oscillate, reverse the tickler coil's connections. Double check your wiring as you go, and the receiver should work without a hitch.

Now a brief operating hint: A small digital frequency counter placed beside your 30-30 receiver makes a perfect wavemeter or "growler" for your vintage station. After tuning in a desired station, simply check the receiver's self-regenerated frequency on your counter. Next key the transmitter (its signal easily captures the frequency counter's front end), and then carefully zero-beat your desired station. Separate transmitting and receiving antennas are suggested here. A single-band dipole is great for transmitting and a 50 foot wire is perfect for receiving (the 30-30 is quite sensitive). I also suggest using this setup on our new 30 meter WARC band. This 10.1 MHz range is terrific for low-power communications, it is open both day and night, and QRM is negligible. Good luck on working the world with your classic 1930s rig and bug!

The Famous 6L6 Transmitter

One of the most undeniably popular homebrew rigs of post-war years was the famous 6L6 transmitter depicted in Figure 8-5. This no-nonsense gem was assembled dozens of ways, ranging from inexpensive versions direct-wired on dimestore breadboards to fancy units enclosed in walnut clock cabinets. If you went first-class in construction, you used an aluminum chassis plus all new parts and added a meter for big-time clout. If you were 3 dB below the poverty level like

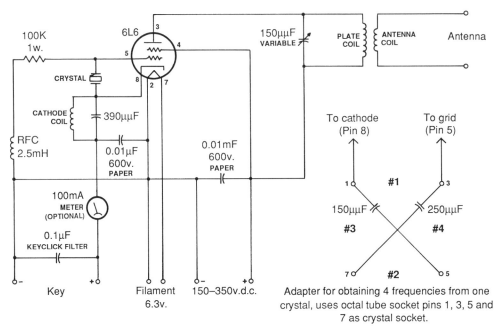

Figure 8-5. Schematic diagram of the famous 6L6 transmitter.

me, however, you built the little transmitter on a wood frame as shown in Figure 8-6. Gaze at this delight with soft glowing vacuum tube and fully exposed tuning condenser (it was still condenser and not capacitor in those days). Now this is a rig that really lights up and does something.

Quick assembly begins by securing a wooden rack like the ones radio stations use to hold 45 rpm records. Cut the rack to half size, add another coat of varnish, reassemble with genuine-era brass screws, and you have two chassis ready to use. Give one to a friend!

Alternately, you can assemble a 3" × 9" × 4½" frame from pine wood. Use ¼ inch wood for the top pieces and space them roughly 1¼ inches apart so tube sockets will fit in the middle and bolt to each side. Mount the 6L6 socket with its key toward the front for easy wiring, and add two Fahnestock clips under right-side screws for antenna connections. Add two more clips or screws with thumbnuts on the left side for key connections, and then proceed to under-frame wiring. (Bell wire works fine here, and it is a sheer joy to use. Doorbell wire is still available from hardware stores, and it is still number 18 solid, but cotton insulation has now been replaced by plastic.)

Look closely at the schematic diagram, and you will notice I added a modern modification of two capacitors to the tube/crystal socket. The crystal thus can be shifted between positions to obtain four slightly different frequencies. If I had only realized that fact 30 years ago!

Plate and antenna coils are wound on a 2 inch diameter jig made by first drawing a circle and then driving seven "7 penny" nails through a thin board. The wire is woven around alternate nails, so every other turn repeats itself. Keep everything symmetrical during winding, and then secure the seven crossover points with kite cord before removing a completed coil from the form. Additional coil data is shown in Figure 8-7.

Use Elmer's glue for mounting four penny-sucker sticks on the wooden frame to hold the coils. Carefully slip the coils in place, wire their connections, and then use more bell wire to wind the cathode

Figure 8-6. The 6L6 transmitter assembled and ready for operation on 30 meters with its mating S-38 receiver. This fun rig is used often at K4TWJ's home, and its collection of DX QSOs grows weekly.

coil on any convenient 1¼ inch diameter round form. (I used a 35mm film canister and wrapped Scotch-brand™ tape around the coil before removing it.) Since this coil is small and light, its pigtail ends are adequate mounting supports.

Several options are possible for powering this little transmitter. If a 150 to 300 volt secondary transformer is available, you are in ham heaven. Otherwise, you can wire two Radio Shack 6 volt transformers "back to back" to acquire both 6 volts for the filament and 135 volts for the plate. Hearty souls might consider direct-rectifying and filtering the 110 volt AC line for plate voltage, but this approach definitely lacks *Good Housekeeping*'s seal of approval. Indeed, 115 volts may be present between your AC/DC receiver's metal cabinet and the transmitter's key if an AC plug is inserted "upside down" in an outlet.

Alternately, you can power the transmitter directly from a series string of twenty 9 volt batteries and a 6 volt lantern battery (borrow all your friends' Radio

BAND	NUMBER OF TURNS	
Meters	Plate coil	Antenna coil
80	20	12
40	12	6
30	10	5
Cathode coil = 5 turns, 1¼" diameter for all three bands.		
NOTE: Doorbell wire used for all coils.		

Figure 8-7. Coil data for the 6L6 transmitter.

Shack free-battery cards). Don't laugh. Batteries are a great source of pure DC, and their life span is good because this fun rig is not intended for main/daily station use. If you cannot find a 6L6 before hamfest time, a 6V6 can be substituted. It is smaller and less glamorous, but it works fine.

Tune-up and operation of this rig are a snap. Apply voltage to the filament; connect a 10 watt house light, large flashlight bulb, or your wattmeter and antenna to the antenna coil; and then apply plate voltage and close the key. Quickly tune the plate condenser for maximum output. If that output seems exceptionally low, try moving and/or turning over the antenna coil on its sucker-stick form. Do not overcouple, ere the transmitter's fine quality be sacrificed.

Next check keying on your modern transceiver and squeeze or warp the cathode coil out of round as necessary to reduce chirps. Tuning the condenser slightly away from maximum output also minimizes chirps, and a milliamp meter in series with the key will indicate input power. Add a .1 mFd condenser at the key's terminals if key clicks are noticeable.

Harmonic suppression with this rig is not outstanding, and output power is only 10 watts, so use a single-band dipole for best results. Fire up this gem on 30 meters and have a ball. Anyone can work DX with a fancy sideband transceiver connected to a big antenna, but doing it with a bare-bones rig built on a wood frame separates the smooth ops from the rough 'gators!

What is a good receiver to use with the famous 6L6 transmitter? A restored-to-new version of your first big-time signal-grabber, naturally! That was not possible in my case, so I combined it with a Hallicrafters S-38. Believe it or not, I worked a number of countries with this setup on 30

Figure 8-8. Is this not an absolutely beautiful classic receiver you would love to operate with a similar-era transmitter and semi-automatic key? John Leary, W9WHM, restored this Hallicrafters SX-73 to better than new, and it is glamorous!

Figure 8-9. Another W9WHM masterpiece. This completely rebuilt Hammarlund Super Pro works even better than it looks. John's hobby is restoring these gems, and "JRL restorations" grace homes of Hammarlund lovers nationwide.

meters and Stateside QSOs were duck soup. Another good receiver you might find at a hamfest is an S-20 Sky Buddy. Think, plan, search fleamarkets, and then purchase and restore your prize.

John Leary, W9WHM, does exactly that today, and his restored classics are collector's items for many thrilled new owners. A couple of John's latest restoration projects are shown in Figures 8-8 and 8-9. If they do not kindle your interest in getting on the air with a classic rig and beautiful old-time bug, check your pulse!

Seriously, however, I trust you enjoyed reading this book on keys and look forward to working you on the air in the near future. It will be much easier to catch me on 30 and 17 meters CW or 20 meters SSB than via mail.

Good luck key collecting, enjoy the pursuit, and have fun using your favorite keys on the air.

Please Remember: Always include a self-addressed, stamped envelope when writing to myself or any other collectors about keys.

Index

Adjusting and using bugs, 41–43
Albright, J. E., 2

Bag Key (Japanese), 34, 37
Bencher, 75
Breedlove, B. H., 26
British army key, 49
Bunnell
    gold bug, 25
    sideswiper, 26

Camelback key, 55
Clapp Eastham key, 45
Codetrol bug, 26, 27
Crystal key, 70

Dow Key
    rotary-yoke bug, 21–22
    tilted-yoke bug, 22–23

Eddystone bug, 27
Electric Specialty bug, 28
Emory's Go Devil, 30

First telegraph setup, 57

G4ZPY
    dual-lever paddle, 78
    pump key, 69, 75
    single-lever paddle, 77
German mouse key, 52–53

Hansen, M., 32
Hartley transmitter, 86
Hi-Mound keys, 50–51, 81–83
Homebrew bugs, 33–34

J-38 key, 47
JRC spy key, 64
Japanese civilian key, 50
Johnson, E. F., 20

Kenpro key, 84
Kent keys, 80–81
Korean key, 84

Landline sounders, 58

Mac-Key, 14–18
Marconi 365, 46
Martin, H. G., 2
McElroy, T. R., 13, 24
Melihan Valiant, 32
Mercury paddle, 84
Miniature keys, 63
Morse, American CW nets, 62
Morse, Samuel F. B., 57
Muse bug, 35–36

Netherlands key, 54

Palm key, 82
Pricing bugs, 40

Refurbishing keys, 40
Rotating wheel bug, 33
Rotoplex by Martin, 12
Russian key, 51

Schurr keys, 78–79
Signal Electric key, 46
Sounder-to-DC converter, 61
Special fingerpieces, 66
Speed-X bug, 18–20
Swallow bug, 31
Swedish key, 53, 82

Vibroplex, 2
    accessories, 71–72
    Champion, 11
    double lever, 3
    identification chart, 13
    Junior, 4

Lightning Bug, 7
  Midget, 6
  Model X, 3
  Number 4 and Blue Racer, 7
  paddles, 73
  Presentation, 4
  Upright, 5
  Zephyr, 12

Western Union, 58

73 Bug, 33